THE MATH MYTH

Also by Andrew Hacker

Higher Education? (with Claudia Dreifus)

Mismatch: The Growing Gulf Between Women and Men

Money: Who Has How Much and Why

Two Nations: Black & White, Separate, Hostile, Unequal

The End of the American Era

Political Theory: Philosophy, Ideology, Science

THE MATH MYTH

AND OTHER STEM DELUSIONS

ANDREW HACKER

THE NEW PRESS

NEW YORK
LONDON

Drawing on page 58 by Whitney Fielding

Requests for permission to reproduce selections from this book should be mailed to: Permissions Department, The New Press, 120 Wall Street, 31st floor, New York, NY 10005.

Published in the United States by The New Press, New York, 2016

Distributed by Perseus Distribution

LIBRARY OF CONGRESS CATALOGING-IN-PUBLICATION DATA

Hacker, Andrew.
The math myth : and other STEM delusions / Andrew Hacker.
pages cm
Includes bibliographical references and index.
ISBN 978-1-62097-068-3 (hardcover : alk. paper) — ISBN 978-1-62097-069-0 (e-book)
1. Mathematics—Study and teaching—Social aspects. 2. Mathematical ability. I. Title.
QA11.2.H35 2016
510.71--dc23

2015022370

The New Press publishes books that promote and enrich public discussion and understanding of the issues vital to our democracy and to a more equitable world. These books are made possible by the enthusiasm of our readers; the support of a committed group of donors, large and small; the collaboration of our many partners in the independent media and the not-for-profit sector; booksellers, who often hand-sell New Press books; librarians; and above all by our authors.

www.thenewpress.com

Book design by Bookbright Media
Composition by Bookbright Media
This book was set in Minion Pro and Gotham

Printed in the United States of America

10 9 8 7 6 5 4 3 2

For
Robert B. Silvers
and
the Memory of
Barbara Epstein

CONTENTS

THE MATH MYTH

1

The "M" in STEM

This century finds America in a struggle to preserve its pride and prestige. Almost daily, evidence accrues that the United States lacks the resources to reign over world affairs. Even minor countries feel free to display their disdain. The years 1900 through 2000 were recognized as America's century. In its first half, we led the world in manufacturing and living standards. In the second, we surpassed in education and military might. But the era ahead holds no comparable promise. Other countries are already matching us in ability, efficiency, and vigor.

For example, we are warned:

- In less than fifteen years, China has moved from 14th place to second place in published research articles.
- General Electric now has most of its R&D personnel outside the United States.
- Only four of the top ten companies receiving United States patents last year were firms based in this country.
- The United States ranks 27th among developed nations in the proportion of undergraduate degrees in science or engineering.

Hence the search for solutions to arrest incipient decline. Piled on my desk are reports from august committees and commissions bearing titles like *The Gathering Storm, Before It's Too Late,* and *Tough Choices or Tough Times.* All call for national revival and renewal. It's revealing that we hear no worries that the United States may lag in literature or the arts, or that no concern is voiced over declining enrollments in philosophy and anthropology. Rather, the focus is on the now well-known acronym STEM, symbolizing competences the coming century will ostensibly require. We must, we are counseled, devote more of our minds, careers, and resources to science, technology, engineering, and mathematics.

Even a decade ago, the Business Roundtable was urging that we "double the number of science, technology, engineering, and mathematics graduates with bachelor's degrees by 2015." We've passed that year, but awards in those fields have barely budged. More recently, a panel appointed by President Barack Obama asked for another ten-year effort, this time to add "one million additional college graduates with degrees in science, technology, engineering, and mathematics." Where the missile race tallied nuclear warheads, now the countdown is STEM diplomas.

THE SOLUTION: AZIMUTHS AND ASYMPTOTES

Indeed, mathematics is the linchpin, heralded as the key to the other three. Thus we're told that if our nation is to stay competitive, on a given morning all four million of our fifteen-year-olds will be studying azimuths and asymptotes. Then, to graduate from high school, they will face tests on radical notations and elliptical equations. All candidates for bachelor's degrees will confront similar hurdles. Mathematics, we are told, will armor our workforce in a merciless world. Its skills, we hear, are foundational to innovation and a lever in the international arena. In an age of high-tech weaponry, azimuths can turn the tide more than human battalions.

On these critical fronts, our young people are no match for their agemates across the globe, with American mathematics scores lagging behind even Estonia's and Slovenia's. A Harvard study calculates that if this shortfall persists, our gross domestic product will drop by 36 percent. The American Diploma Project reports that proficiency in algebra will be needed in 62 percent of new jobs in the decades ahead. We are further warned that there's already a shortage of graduates with STEM skills. As a result, vital work is being sent abroad, or credentialed immigrants are enlisted to fill the vacancies. If coming generations want a quality of life they feel is their due, they must be prepared to master what many find the most difficult of all disciplines.

Back in 1841, a Scotsman named Charles Mackay (it's his spelling) published a book called *Extraordinary Popular Delusions and the Madness of Crowds*. He showed how hoaxes, frauds, and hallucinations come in varied guises, from stalking witches to marching off to wars. In more sophisticated times, delusions must show a surface plausibility if they are to ensnare an ostensibly educated populace.

Among our era's delusions are the powers ascribed to mathematics, spurred by a desperate faith in skills abbreviated by the STEM acronym. Together, they have animated a major mythology of our time. Like all myths, they start with a modicum of truth and can be beguiling on first reading. In the chapters ahead, I will show why these beliefs, even when sincerely held, are wholly or largely wrong, lacking in factual support, and usually based on wishful logic. More consequential, these illusions and delusions are already taking a heavy toll on this country, most markedly on the humane spirit that has made America exciting and unique.

FEARED AND REVERED

The Math Myth began nearly twenty years ago, when I started making notes, conducting interviews, and collating files. For much of the time, it had an intermittent schedule, competing with other projects. This changed in 2012, when editors at the *New York Times* heard about what I was up to and asked me to write an opinion article on mathematics. Responses poured in, at close to record levels, which told me it was time to finish the book. So I did, and here it is.

I mention the early beginning because even with the passage of years much of the terrain remains unchanged. Interviews conducted at the outset remain fully relevant today, as are facts and figures I amassed. If anything, myths about mathematics—the central subject of this book—have become more entrenched. This is why I believe that *The Math Myth* is needed. This country has problems. But more mathematics isn't one of the solutions.

Other books of mine have ranged broadly, from race and wealth to corporate power and the gulf between the sexes. I've also sojourned in philosophy, writing on titans like Plato, Hegel, and Rousseau. So why now mathematics? My answer is that I've found it an absorbing example of how a society can cling to policies and practices that serve no rational purpose. They persist because they become embedded, usually bolstered by those who benefit. Nor are the issues entirely academic. Making mathematics a barrier ends up suppressing opportunities, stifling creativity, and denying society a wealth of varied talents.

Although I have taught in a department of mathematics at a respected college, strictly speaking, I'm not a mathematician, in that I have no degrees in the discipline. Still, I can admit to being a social scientist, which has always had a quantitative side. I also have a fair reputation for being agile with numbers. But this is not a *mathematics book*, in the sense of being a textbook or a volume on the beauties of topology. Rather, it is *about* mathematics, as an ideology, an industry, even a secular religion.

As a long-gone wag once put it, mathematics is both feared and revered. Feared by those who recall it as their worst academic subject, if not a class that blighted their day. Mathematics is also revered, as an inspiring achievement, all the more esteemed because so much of it is beyond our ability to grasp. Amid this aura of awe, it's easy to argue that even more of it should be taught and learned. Yet on the other side is the obdurate fact that of the millions of high school and college students subjected to mathematics, distressing proportions are slated to fail.

This book asks a seemingly ingenuous question: why do we impose so prolonged a sequence of a single discipline, with no alternatives or exemptions? Given our skeptical age, I find it curious that hardly anyone has asked.

PASSING PASCAL

The mathematics regimen is already well entrenched. Under current expectations, every young American will study geometry, trigonometry, plus two years of algebra, with talk of adding calculus to the menu. These requirements have consequences. Currently, one in five of our young people does not finish high school, a dismal rate compared with other developed countries. Of those who manage to graduate and decide on college, close to half will leave without a degree. At both levels failure to pass mandated mathematics courses is the chief *academic* reason they do not finish. (Notice the italics. There are other causes, including prison and pregnancies.)

The usual responses reflect this country's can-do spirit. Calls are heard to return to rigor and end feel-good nostrums. Our goal should be to make the entire nation mathematically adept, starting early with those of school age. Just as wars were declared on poverty and drugs, so our classrooms must be viewed as battlegrounds, armed with more qualified teachers and stringent curriculums.

Clarions like these gave birth to the Common Core State Standards, which at this writing hold sway in more than forty states. On coordinated dates, all public school students in the states will take the same or parallel tests on specified subjects. The most decisive tests—gauged by the possibility of failure—will be in mathematics, where pupils will confront questions like these:

Use the properties of exponents to interpret expressions of exponential functions. For example, identify percent rate of change in functions such as $y = (1.02)^t$, $y = (0.97)^t$, $y = (1.01)^{12t}$, $y = (1.2)^{t/10}$, and classify them as representing exponential growth or decay.

Know and apply the Binomial Theorem for the expansion of $(x + y)^n$ in powers of x and y for a positive integer n, where x and y are any numbers, with coefficients determined for example by Pascal's Triangle.

To believe that such equations are a solution is yet another instance of self-delusion. I will propose that we can deploy our material and human capital in better ways. Our goal should be to keep our young people in high school and later in college, where they can discover and develop their talents At this point, we're telling them that they must unravel reentrant angles and irrational numbers if they want a high school diploma and a bachelor's degree. I'll be offering alternatives.

"THE GREAT BOOK OF NATURE"

In no way is this book "anti-mathematics," if such a stance is possible. I would be elated if everyone understood what mathematics is and does, along with the breadth and depth of its achievements. Sadly, few mathematicians seek to evoke this appreciation, whether in their classrooms or to wider audiences. I will argue that if mathematics is to join the liberal arts, it needs to meet the rest of the world halfway.

I'd also like everyone to appreciate how mathematics undergirds our lives. Peter Braunfeld, on the faculty at the University of Illinois, once remarked to me that "our civilization would collapse without mathematics." Its models design racing cars, tell retailers how many sweaters they need for a holiday season, conjure crowd scenes in fantasy movies, and guide hotels on rates that will fill as many rooms as possible. I'd also like everyone to know how trigonometric functions enabled the Wright brothers to keep their plane aloft for a historic minute, just as calculus now allows a 450-ton jetliner to soar nonstop from Hong Kong to New York. Courtesy of equations we never see, our lives are safer and more varied and interesting. But this isn't being taught by mathematicians, because it doesn't align with their rigid syllabus. Nor are they inclined to allow anyone else to tell this story. As a result, one of the most exciting sagas of our time remains unknown and untold.

So, yes, I honor mathematics as an awesome intellectual enterprise. I will gladly have my tax dollars go for research on Goldbach's Conjecture and parametric cycloids. Not the least of my concerns about STEM is that it casts mathematics largely as an arm of technology, in a global competition gauged to gross domestic product, military might, and electronic surveillance. I would like all liberal arts students, indeed everyone, to know what Galileo meant when he called mathematics "The Great Book of Nature." Or why Isaac Newton subtitled his *Principia Mathematica* a study in "Natural Philosophy." I would urge schools and colleges to draw on perspectives like these and accept them as coequal to precalculus and trigonometry.

THE IMPORTANCE OF ARITHMETIC

There are some other observations I'd like to include before this book gets under way. One is to highlight the difference between mathematics and arithmetic, if only because the terms tend to get conflated. For instance, we speak of "mathematics scores" of third-graders, when all they were tested on was plain-vanilla arithmetic. Mathematics basically begins in high school, extends from geometry through calculus, and ultimately soars to the ethereal pursuits of specialists and scholars. Arithmetic is addition, subtraction, multiplication, and division, followed by fractions, decimals, percentages, and ratios, and the statistics we encounter in our everyday lives. All of us should get arithmetic under our belts in elementary school.

At this point, the challenge is not to immerse more people in more mathematics. Rather, it is that many, if not most, young people and adults are insufficiently agile in arithmetic. Of course, they can add and subtract. But a 2013 study of adults in twenty-three countries found the United States third from the bottom in ostensibly simple tasks like using odometer readings to submit an expense report. If that's a cause for worry—and I believe it is—it needs to be remedied on its own terrain. There is ample evidence that mathematics as it is currently taught does not improve quantitative literacy or quantitative reasoning, or facility with the figures that inform and organize our lives. Even if most American adults once studied algebra, geometry, and phases of calculus, it hasn't enhanced their numerical competence.

What's needed is what I'll be calling *adult arithmetic,* or what John Allen Paulos has termed *numeracy.* Nor does this mean revisiting fourth grade. Numeracy can and should be taught at a demanding level. I'll give some examples in my closing chapter, where I'll show how arithmetic is all that's needed to interpret charts in the *Wall Street Journal* or graphs in *The Economist,* as well as public documents and corporate reports.

The editors at the *New York Times* chose to title my article "Is Algebra Necessary?," implying I thought it was not. (For my part, I would have phrased the question "How Much Mathematics Is Too Much?") I admit that in this book I occasionally use "algebra" as a surrogate for the full mathematics sequence. That noted, I want to affirm that basic algebra is definitely necessary for everyone. I use it every day myself. If a twenty-muffin recipe specifies thirty-five ounces of flour and we want only thirteen muffins, how much flour (x) do we need? The equation we jot down—20 is to 13 as 35 is to x—is elementary algebra, or simply "solving for x." True, this involves only multiplication and division. Still, every teenager and adult should have this skill and understand the reasoning behind it.

The current regimen expects students to master not only *x*-equations, but also associative properties, squared binomials, and prime factorization. Anthony Carnevale and Donna Desrochers once remarked that there can be such a thing as "too much mathematics." Those words deserve respectful pondering.

OPTIONS AND ALTERNATIVES

Arithmetic has always been a required subject, as it certainly should. But ought mathematics be made optional? Should ninth-graders be allowed to forgo geometry because they've heard it's hard? Paul Lockhart, a virtuoso mathematics teacher at St. Ann's School in Brooklyn, thinks so. "There is no more reliable way to kill enthusiasm and interest in a subject than to make it a mandatory part of the school curriculum," he says of his chosen field. Part of me agrees with him. As a college professor, I teach mostly electives, so I'm spared from facing sullen conscripts. Still, once Lockhart's door is opened, there's the problem that students might also ask to opt out of science, literature, and history, or physical education.

Yes, I want changes to come, but not solely on the votes of fourteen-year-olds. Among its many audiences, this book is addressed to adults who preside over our schools, and parents who support them. I want to urge them to consider alternatives to the current mathematics syllabus. Growing numbers of reputable colleges now accept applications without asking for reports from the SAT or ACT, which says they feel no need to scrutinize mathematics scores. Moreover, the students who decide not to submit them have been found to end up doing just as well in their courses. In this vein, I urge innovative school systems to experiment with allowing several of their high schools to offer arts or humanities diplomas, perhaps starting in the tenth or eleventh grade, akin to options many European systems offer. Such sequences might also create courses in statistics or quantitative reasoning, in lieu of conventional mathematics.

ACROSS A SOCIAL SPECTRUM

When I propose allowing alternatives, it is not to spare young people from lessons they'd rather avoid. A common response is that varying the syllabus will result in "dumbing down" the curriculum. In fact, my proposals don't constitute a downward move at all. On the contrary, they call for smartening up. I will show that versatility with figures is as exacting as trigonometry or geometry, while becoming deft with statistics calls for as much reflective reasoning as advanced algebra.

There is a related concern that offering alternatives will essentially install a subordinate track that would be occupied largely by pupils from already-marginalized groups. Young people even now behind the pack would be immured there, or so it is feared. To deny them the skills associated with mathematics, this argument goes, would bar them from the culture and careers that will exemplify our coming century. This view has inspired Robert Moses's Algebra Project and is seconded by Eric Cooper of the National Urban Alliance for Effective Education. Both basically say that since society has erected a set of hoops, it's best to learn how to jump through them. Ambitious immigrant families have long instilled this acquiescence in their children.

I might echo this opinion if it had firmer ground. As subsequent chapters will show, there is no evidence that academic mathematics, at least not the kind students are being urged to learn, will be the lingua franca of the future. New talents of many sorts will surely be needed. But first to require that all show proficiency in parabolic geometry will actually hinder the emergence of strengths not based on equations.

Moreover, advanced algebra is as much a stumbling block for students in spacious suburbs and upscale neighborhoods as it is in rural counties and inner cities. I'll add only that in the former, the stumbling is covered over by private tutors and coaching courses. Many of the letters I receive are from professional parents, whose sons' and daughters' lives have been mangled by mathematics barriers. Nor are they, or others across town, looking for easier options. Young people across a social spectrum have potentialities in a host of exacting fields. Yet the SAT, ACT, and the Common Core, with their unyielding stress on trigonometry, precalculus, and advanced algebra, have created arbitrary and intractable barriers for students whose aptitudes lie outside of mathematics.

ANOTHER PATH

My concern, as an educator and a citizen, is over the precedence accorded to science, technology, engineering, and mathematics in expanding spheres of our society. I am not saying we're at risk of becoming a nation of robots or nerds or geeks. Still, I worry when we're told that mathematics' special mode of reasoning should take precedence in analytic endeavors. As when a University of California professor asserts that in our new world "math skills are more important than literary skills." I use numbers more than most people, but I find this assertion to be ominous. I'm not sure I'm ready for conversations couched in equations.

Having opened this chapter with musings on the hegemony of STEM, I'd like to close by proposing another acronym: PATH. In my rendering, it stands for Philosophy, Art, Theology, History. (Or try your own: Poetry, Anthropology, Theater, Humanities?) And thus this declamation.

> *We live in critical times, because we are falling behind our competitors in PATH pursuits: philosophy, art, theology, and history. If our nation is to retain its moral and cultural stature, we must underwrite a million more careers in PATH spheres every year. If we do not, we may continue to lead in affluence, but we will decline as a civilization.*

2

A Harsh and Senseless Hurdle

PLUS C'EST LA MEME CHOSE

Back in 2007, John Merrow wrote an article for the *New York Times* that focused on a student at LaGuardia Community College named Krystal Jenkins, who aspired to become a veterinary technician. Seven years later, in 2014, Ginia Bellafante, writing for the same newspaper, visited the same college. Her article's center was Vladimir de Jesus, who hoped to major in studio art. The stories were strikingly similar. They told of students denied a chance for further education due to crushing mathematics requirements.

"I love animals," Krystal Jenkins said. Her fetching way with them spurred her desire to become a veterinary technician. But before she was allowed into even introductory classes, her college demanded that she pass a course in linear and quadratic equations. Sadly, Krystal failed it twice, and was told she couldn't take it again. "It all came crashing down," she said, as she left the college for good. Not a single veterinarian or technician with whom I've spoken could recall a need for that level of algebra. Quite obviously, numbers figure in prescriptions, inoculations, and treatments. But accurate arithmetic suffices.

Vladimir de Jesus had been allowed a third try in a similar LaGuardia class, featuring assignments like the cosine of pi over two. But he found it "a stainless-steel wall and there's no way up it, around it, or under it." Nor was he alone; 40 percent of his classmates also failed. Vladimir has joined Krystal in a growing population of involuntary alumni, the victims of a one-size-for-all ideology. He has turned to freelance tattooing.

DISMAYING STATISTICS

In the percentage of its young people who finish high school, the United States ranked twenty-second out of thirty developed countries surveyed by the Organization for Economic Cooperation and Development in 2013, behind Hungary, Slovenia, and Chile. The United States did a bit better on college completion: it was twelfth amid thirty-two nations, this time after Luxembourg, Israel, and New Zealand. We have more colleges per capita, but fewer of our students stay to finish a full program.

In our high schools, one in five of our ninth-graders doesn't make it to a diploma. In New Mexico and Georgia, 28 percent aren't in the procession. In Nevada, it's 29 percent. That means that each year a million American teenagers start life without a basic national credential. Of those who do graduate and enter college, only a little over half—56 percent—emerge with a bachelor's degree. America's educational highway is littered with dropouts at every mile, a human roadkill that doesn't have to happen.

A lot of reasons have been cited for these shortfalls in our education system. We're a far larger country than Portugal. We're more complicated than Iceland. We have higher pregnancy rates among teens than many developed countries. We are extremely punitive and put more of our young people in prison. Our poverty rate is the highest in the developed world.

I believe these factors may all play a role, though I also feel it is within our power to address and remedy most of them. That said, this book will explore another, *academic* reason our efforts at universal education fall so conspicuously short: our insistence on heedless and needless mathematics requirements. Research by Lynn Arthur Steen of St. Olaf's College shows that "mathematics is the academic subject that students most often fail." Jo Boaler at Stanford University goes a step further: "currently, more than half of all U.S. students fail mathematics." Nor should this surprise us. History, literature, and biology all touch base with realities we know. Compared with other subjects, mathematics represents an alien world, an enigmatic orbit of abstractions. To be sure, most pupils eventually squeak through. But the pass rate for mathematics is the lowest of all departments and disciplines.

Still, in the name of college preparation and academic rigor, each year sees even more mathematics being imposed. As recently as 1982, only 55 percent of high school graduates had a course in algebra, and 47 percent had taken geometry. Today, 88 percent who finish have had a geometry course, and 76 percent have had two years of algebra.

HIGH SCHOOL: ALGEBRA AND ATTRITION

Shirley Bagwell, a high school mathematics teacher in Tennessee—who has written articles with titles such as "When Barbie Drops Algebra" and "Is Algebra Hurting America?"—warns that "to expect all students to master algebra will cause more students to drop out." Most who get through, she adds, "will avoid mathematics forever and remember the subject as a nightmare." She is seconded by Teresa George, a veteran Arkansas teacher: "Some students are never going to pass algebra, and then you've lost them." The late Gerald Bracey, another classroom veteran, adds that "forcing everyone to take algebra is more likely to turn kids off mathematics and even off school altogether." Extensive research supports what they are saying. (Here and elsewhere, I and others use algebra as a shorthand for the usual mathematics menu in secondary education. In fact, geometric parabolics can be just as daunting as algebraic vectors.)

A report entitled "Locating the Dropout Crisis," by Robert Balfanz and Nettie Legters of Johns Hopkins University, found that "failing ninth-grade algebra is the reason many students are left back in ninth grade, which in turn is the greatest risk factor for dropping out." A study of Los Angeles schools, supervised by David Silver of the University of California, was more precise: "on average, 65 percent of students in any given Algebra I class in the district will fail." Geometry fared somewhat better, with only half—51—percent failing.

Nor is it only teachers who hand out failing grades. Before the advent of the Common Core, several states had their own comprehensive tests. Prodded by the misnamed No Child Left Behind law, they installed "exit" exams, which all students had to pass to secure a diploma, pretty much ensuring that at least some children would be left behind. By 2013, at least nineteen states had mandated such exams and had reported results. Their failure rates for mathematics were arresting. In Minnesota, 43 percent of students taking the mathematics test didn't pass. In Nevada, it was 57 percent; Washington, 61 percent; Arizona, 64 percent. And these are students who, at least until the test, had made it to their senior year. In all but one of the nineteen states, the highest failures were in mathematics, not some other subject. This happened not because that many young people were indolent or indifferent. Rather, they hadn't mastered equations even educated parents have forgotten.

As of this writing, most states have signed on to the Common Core, with its plans for uniform tests and parallel scoring systems. Their next step is to decide if they want to require specific Core scores for high school graduation. For example, will identical grades for "passing" or "proficient" be used nationally? Since the Core's "standards" include advanced algebra, a state like Alabama will have to choose whether to use the same passing score as, say, North Dakota. If Alabama does, it may find a majority of its high school seniors leaving without diplomas.

Each year also sees more states requiring a second year of algebra. Often the impetus comes from legislators and business groups, hardly any of whom have a notion of what is taught in actual classrooms. Joseph Rosenstein, a Rutgers University mathematics professor, can find no rationale for imposing such specialized concepts on everyone: "It is hard to make the case that topics like complex numbers, rational exponents, systems of linear inequalities, and inverse functions are needed by all students." Rosenstein asks these lawmakers and executives, "When was the last time you needed to factor trinomials?"

We know that students' grades and scores tend to correlate with their parents' social and economic status. (There's at least one exception to this rule, which I'll touch on later, with respect to ethnicity.) As we might expect, more pupils who fail are from low-income homes. But that's not the critical issue with mathematics. It is a hurdle for all kinds of students, both disadvantaged and affluent, as well as from all ethnic origins. In New Mexico's mathematics exams, 43 percent of white pupils fell below proficient, as did 39 percent in Tennessee. Nor is this surprising. We all know professional families where one daughter is a mathematics whiz, while her sister turns semi-suicidal over geometry. Still, with intensive (and often expensive) coaching, the second child may manage a passing grade.

Enter the coaching industry, a mainstay when academic opportunities hinge on multiple-choice points. A 2015 estimate found corporations like Kaplan and Princeton Review taking in $7 billion for classes and personal sessions, with at least another $3 billion going to freelancers. Needless to say, the bulk of test preparation is for mathematics. It's hard to see a demand for social studies tutorials.

Colleen Oppenzato, a tutor who works from her Brooklyn apartment, told me she spends most of her time with students explaining how the test is structured and teaching test-taking techniques, rather than the mathematics itself. One such technique is "back solving," where the student starts by examining the alternative answers rather than the actual question. "With shrewd tutoring," she told me," someone knowing no mathematics at all could get a four hundred on the SAT." True, that's not an auspicious score. But it's an indication of how far test-taking technique can affect performance.

A survey by a local newspaper in suburban Pelham, just outside New York City, found that over half of the families were paying for such aid. And this was for students already benefiting from an elite school system, with a well-credentialed faculty. It adds up to an admission that even well-performing schools cannot fully prepare their pupils for the mathematics marathon, which each year moves the starting line to an earlier age. If suburbs such as Pelham, with a median household income of $114,444 (about double the national figure), find a need for tutoring, imagine the deficits in the rest of the country.

Mathematics hurdles underscore and contribute to the country's social divide. On one side are families who can afford to reside in well-endowed districts with smaller classes and attentive teachers, supplemented by private classes and coaching. Kaplan and Princeton Review ask about $800 for ten small-group Saturday-morning sessions. On the other side, a Manhattan tutoring service wants $700 an hour for one-on-one tutoring, although for that they come to your home.

CLOSING COLLEGE DOORS

Whether everyone should attend college has always been a contentious question. (Much turns on when; perhaps not everyone should start at eighteen.) If we want to keep opportunities open, admissions standards become an issue. So in the name of rigor, most colleges now want all applicants to arrive with at least three years of mathematics, including those rational exponents and linear inequalities taught in the second year of algebra. Most colleges also require respectable scores on the SAT or ACT, which each include problems involving complex numbers and inverse functions.

The twenty-three campuses of the California State University system, stretching from Fresno to Stanislaus, aren't Stanford or

Berkeley. Even so, to be considered by any of them, students need to have mastered the full mathematics menu, including two years of algebra. So students who show promise in art history or post-modern criticism won't even have their applications opened if they faltered in geometry. The consequences are dismaying. In 2012, only 38 percent of California's high school graduates—who actually won diplomas—were deemed eligible for branches of Cal State. Graduates classified as Caucasian did somewhat better than the overall average, but not by much. Only 45 percent had transcripts deemed worth considering by their state's public colleges. Anthony Carnevale and Donna Desrochers ask, "are mathematics courses creating artificial barriers to college entry?" By now, the answer is obvious.

We often hear that rejected applicants can start at two-year colleges and continue with their education if they do well there. This too is largely a myth. A 2013 study by the Century Foundation indeed found that over 80 percent of students starting in community colleges *wanted* to proceed to at least a bachelor's degree. Sadly, a follow-up six years later found that only 12 percent had actually been able to fulfill that aspiration. What deterred them? By this time, the reason should be evident. Once admitted to two-year schools, they find themselves in a mathematics morass.

In May of 2014, Paul Tough wrote an incisive analysis, "Who Gets to Graduate?" His first answer was that it wasn't community college students. On arrival, fully two-thirds of them are consigned to remedial mathematics classes, for which they get no credit, but which they must pass in order to enroll in any other courses. It isn't that they can't do long division, construct ratios, or interpret statistical tables. No, the colleges want them proficient up to trigonometry, before they can enter programs in commercial art or cosmetology.

At Pennsylvania's Montgomery County Community College, half of the students taking required mathematics courses end with failing grades. A study of twenty-seven two-year colleges nationwide found that less than a quarter of the students pursuing a mathematics requirement had fulfilled it three years later. "There are students taking these courses three, four, five times," said Barbara Bonham of Appalachian State University. Even if some ultimately pass, she adds, "many drop out."

Tennessee provides another dismal case, where over 70 percent of its college freshmen are consigned to remedial mathematics sections. They apparently find the experience so dispiriting that only five percent of them graduate on schedule. In 2013, a nonprofit research group called Complete College America issued a report called *Remediation: Higher Education's Road to Nowhere.* Its authors confirmed that "most students are placed on algebra pathways." Yet, it suggested, a little thought would show that, instead of algebra, "statistics or quantitative mathematics would be most appropriate to prepare them for their chosen programs of study and careers." Nor is this advice just for occupational preparation. Many two-year students opt for liberal arts fields, often to enhance their lives or become better citizens. Courses in statistics geared to their interests and needs would suit them better than algebra. Thus far this sensible proposal has found hardly any takers.*

Another study, this one by the National Center on Education and the Economy, found that "many community college students are denied a certificate or diploma, because they have failed in a mathematics course irrelevant to the work these students plan to do or the courses they need to take." Marc Tucker, the report's principal author, believes that algebra "is being used much as Latin was used a century ago, as a screen to keep the unwanted out of college." Much of the blame rests with two-year faculties and administrators. Often on the defensive, and anxious to elevate their status, they try to show how rigorous they are by piling

---

*Is remedial work needed? It often is. In my introductory political science classes, I encounter far too many students who cannot write coherently. I would like to think this is being remedied in composition courses, since without that competence they can't do college-level work. But it's quite another thing to say that *all* undergraduates need advanced algebra to proceed toward their degrees.

on mathematics. They may get some professional points. But it's their students who pay the price, often being forced out of education entirely.

As will be shown in the next chapter, the kinds of mathematics taught in classrooms have little or no relevance, even in technical fields like electronic drafting. Lynn Arthur Steen is a rare mathematics professor willing to say this candidly. "What prospective employees lack is not calculus or college algebra," he points out, "but more basic quantitative skills that could be taught in high school."

COLLEGE DREAMS ARE DASHED

For over a century, the United States led the world in opening higher education to increasing numbers of its citizens. No other nation enacted anything like the Morrill Act of 1863 and the G.I. Bill of 1944 to put degrees within the reach of so many. Currently, over 60 percent of Americans give a two- or four-year college a try, if only for a semester. But the less heartening side, as previously noted, is that of the 2.5 million who sign up each September, almost half don't finish. And they are a rather different set than those who didn't complete high school. After all, they stuck it out for a high school diploma, demonstrating their capacity for academic work.

The greatest attrition is in the first year of college. Here, again, the primary academic cause is a mathematics requirement, all too often demanded of all students regardless of their prospective major. A City University of New York study of its mandated algebra course found that 57 percent of the students failed. In one of its member colleges, 72 percent didn't pass. The report calculated that the failure rate in mathematics was two and a half times greater than for the rest of the curriculum taken together. Here was its depressing conclusion: "failing mathematics affects retention more than any other academic factor." It would be nice if that had sounded a wake-up call, but as of this writing those requirements are still in place.

The most extensive data are from the Institute on Postsecondary Education, which examined transcripts containing over 265,000 grades at a cross section of colleges. For each course, it added up its failures, withdrawals, and incompletes. By this time, we shouldn't be surprised to hear that nonpassing in mathematics was two to three times that for all other courses.

Almost a dozen countries now have a greater percentage of their population completing college than the United States does. Many of these countries require mathematics only for fields where it is rationally warranted. The United States has just as much nascent talent, but has instituted irrelevant requirements that bar many talented young people from higher education.

Most undergraduates don't want to take mathematics, and few in freshman mathematics courses are there voluntarily. Compounding the problem, the teachers they meet in introductory sections are usually adjuncts or graduate assistants. Even experienced adjuncts tend to be overworked, with cramped office space, and are often rushing to another job. Teaching assistants are seldom counseled or supervised; more than a few nowadays are newly arrived in this country and made to teach as a condition of receiving a stipend. Suzanne Wilson of Michigan State University concluded that most mathematics faculties are not held accountable. Indeed, she found that when "a student fails a course, no mathematician is obliged to go back and help her learn what she did not understand."

Other fields try to make their first-year offerings interesting and appealing: no one is required to take anthropology. So if that department wants to recruit students to the discipline, it must make its introductory offerings attractive. "Few freshmen have even heard of anthropology," Kevin Birth, its chairman at Queens College, told me. "While we in no way dumb it down, we show how our discipline can enhance their understanding of the world." That's not the attitude in mathematics. Some departments boast of "weeding out" pupils who don't come with a commitment to the subject. Or they divide freshmen into "plums" (the few deserving of faculty attention) and "prunes" (the majority, to be discarded). They can maintain this posture no matter what they do, because each year brings a new involuntary intake to bolster the departmental budget.

IVIED TOWERS

The nation's top-tier colleges revel in how many students they reject. In a recent year, 29,610 applications poured in to Yale. Given all these folders, where does the admissions process start? The answer is to rank them by SAT scores, putting those with the 1600s—perfect scores in the verbal and mathematics sections—at the top of the pile. (Selective schools generally ignore the "writing" section.) Today, the elite colleges demand top ratings in *both* parts of the SAT. Harvard, Princeton, and Yale expect three-quarters of the students they accept to achieve a score of at least 700 in mathematics, a height reached by only nine percent of all men nationwide and only four of every hundred women. Stanford, Duke, and Dartmouth aren't far behind, looking for a score of at least 680. Scores that high are also rare, achieved by only twelve percent of men and six percent of women. One result of such filters, to be explored more fully later, is that more women who are wholly qualified apart from mathematics find themselves being turned down.

True, a quarter of those who are accepted are below the 700 cutoff. But it's a safe bet that many of them are sought-after soccer players, or offspring of donors or alumni. Others are admitted under affirmative action, which includes not only ethnic backgrounds, but candidates from underrepresented states like Alaska and Montana. I am not suggesting that there's anything wrong with being a top mathematics scorer. But Dartmouth and Duke aren't Cal Tech and MIT. They are primarily liberal arts colleges, ostensibly committed to the breadth and depth of learning. Yet for three-quarters of the classes they admit, a very high mathematics showing is required, even for those planning to major in philosophy, classics, or modern dance, all of which are majors Ivy League and other elite colleges offer.

Even if technology and science are destined to loom larger than in the past, a lot more will be wanted and needed in our society. It's unrealistic to expect that all the talents we want and need will always be found in tandem with mathematics. As a society, we had best be careful that we are not constricting—not to say contorting—our conception of excellence.

VOICES

My daughter is a published playwright and a finalist in a New York festival. After being accepted by a college theater program, she was denied entry because of her poor math scores.

We have been discouraging people who would have made really good doctors and lawyers (and welders) by making them prove themselves at irrelevant math tasks.

A student of mine's only desire was to become a sports writer. He must have taken algebra at least five times. To deny him to the world of journalism because of algebra has to be some kind of foolishness.

I'm one of those people who faced the brick wall of algebra during my aborted attempt at a bachelor's degree. I now write, produce, and direct TV programs, and have won an Emmy award. Without a grasp of algebra, I still manage to keep tidy books for my production business.

VOICES

Although I already had an Ivy degree, I was required to take an algebra course because they claimed I would need it in social work.

My son's college career ended because of algebra. He is now an accomplished photographer. But the fact that he is the only one in our family without a degree continues to haunt him.

If I were the same student I was years ago, required to take algebra in order to graduate from college, I would be one of the so-called dropouts.

I'm still steaming that an algebra requirement prevented me from taking a high school seminar in American history.

Will Plumbers Need Polynomials?

Here's what we're being told:

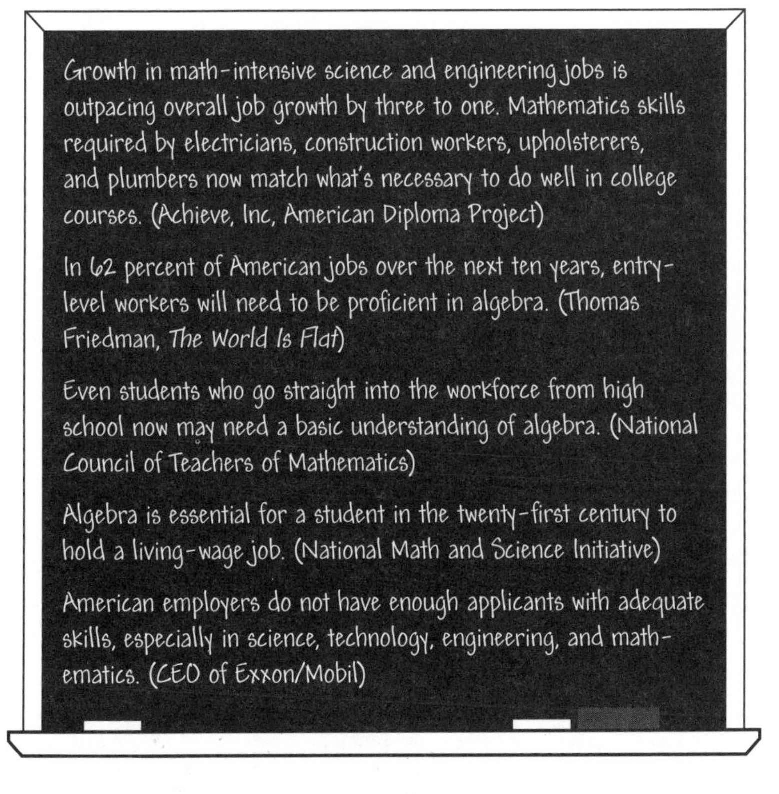

There's one problem. Despite their august auspices and air of authority, all of these statements are wrong. Ours is obviously a high-tech age, whose momentum will certainly continue and in all likelihood increase. For my own part, I readily grant that mathematical equations are crucial to innovations on which we rely. But we're hearing something more. It's that to meet the demands of a new era, we must *all* become expert in what until now has been a specialized sphere.

What's being proposed is radical. While it is slated to start with youngsters in our schools, its ramifications will be felt across all of our society. Already installed is the Common Core, which will require every young person in the nation to master advanced algebra regardless of his or her interests or aspirations. The reasoning for this purports to be practical; even midlevel occupations, we are being told, will have tasks involving trinomials. But let's look at the actual employment picture, as it now appears and where it is heading.

On this, the Bureau of Labor Statistics is generally accepted as the most reliable source. The BLS keeps count of thousands of occupations in the United States, from architects to zoologists. (In 2012, it reported there were some 107,400 of the former, 20,100 of the latter.) After that, the agency analyzes social and economic trends, from which it forecasts how many jobs will be created in various sectors of the workforce over the coming decade. Its projections appear every two years in a comprehensive document called the *Occupational Outlook Handbook*, which is fully available online.

I delved into the 2014–2015 edition to see what it said about employment overall, and about mathematics-related occupations specifically. The *Handbook* said that the total number of jobs in the country in 2012 was 145,355,800, and predicted that this number would rise to 160,983,700 ten years later. So adding 15,627,900 new paid positions would make for a 10.8 percent rise during the decade.

The BLS doesn't forecast how many jobs will have an explicit need for mathematics, nor does it estimate a figure for STEM jobs taken together, if only because there's no consensus on which positions should be included. (Perhaps architects, but zoologists?) The best estimates I've seen are in a still relevant 2008 analysis conducted by the respected Georgetown University Center on Education and the Workforce. In that year, the Georgetown analysis found that 7.3 million men and women held positions calling for mathematics-based skills, which came to five percent of all employed adults. In fact, these figures are somewhat generous, since the Georgetown group gave STEM status to occupations like school psychologists, landscape architects, and archaeologists. They projected that the STEM share of total employment would rise modestly, from 5.0 percent of all jobs in 2008, to 5.3 percent in 2018: a three-tenths of a point increase over ten years.

Returning to the BLS employment numbers, we discover that some STEM occupations will be expanding only modestly compared with the decade's overall increase of 10.8 percent. During these ten years, in the nation as a whole, the BLS predicts a need for 5,400 additional chemists and materials scientists, 2,400 physicists, and only 800 new mathematicians. Taken together, this adds up to fewer than 900 new spots per year for these types of scientists. Indeed, the 4.4 percent growth for chemists is not even half the overall growth for hiring. The percentage for mathematicians is a bit better, at least if they want to work for the National Security Agency, which will likely be their most prominent employer.

NOT ENOUGH ENGINEERS?

The BLS's analysis for engineers saw their numbers going from 1,452,400 in 2012 to 1,582,800 in 2022. All told, these 130,400 newcomers make up a nine percent rise, below national new hirings of 10.8 percent. This means there will be *less* demand for engineers, compared with other occupations. How can this possibly be? We've long been told engineers are the spinal cord of technology, whether building roads and bridges, fabricating mainframes, or designing software.

Of the fifteen engineering areas studied by the BLS, nine will actually have fewer people working by 2022. These net losers include stalwarts like mechanical and electrical engineering, as well as the aerospace and nuclear specialties. Moreover, some of the gains will be markedly modest. Health and safety engineering will add only 260 places per year, while mining will average 100 new annual hirings.

For my part, I believe that the United States *should* have more people doing certain kinds of jobs—say, family physicians or investigative journalists. After all, the work they do advances the common good. But that's mainly a moral statement, much like saying it would be nice if we had more faithful spouses or considerate drivers. So we would do well to recall what we learned in Economics 101. New jobs appear only when money is made available for their wages. Private firms create openings when they feel that the newly employed workers will augment their bottom line. Government and nonbusiness organizations embark on hiring only if they get the funding from taxes or other sources. Thus a company like Disney will budget salaries for more mathematicians if it decides to mount new animation projects. In a similar vein, NASA can add aerospace engineers only if its appropriations are increased.

I recall a conversation concerning China I had with Joseph Stiglitz, the Nobel laureate in economics. That country is graduating hundreds of thousands of new engineers every year, and we are being urged to match or at least get nearer to their figure. Suppose we did, he said, adding, "I don't think that we would be able to integrate into our economy anywhere near the number of engineers that they have." Simply stated, China has salaried jobs for its technical graduates due to public funds invested in infrastructure projects, while its trade surpluses underwrite a burgeoning industrial section. "The United States looked much that way a century ago," Stiglitz added. "It's naive to think we can re-create that age."

Currently, there isn't a cache of cash waiting to provide wages for new infusions of STEM graduates. Sometimes it's the other way around. A Florida community college launched a program in laser photonics, because a nearby Northrop Grumman facility said it needed technicians with that skill. Sadly, the Pentagon cut the program, not only leading to layoffs, but leaving people with training but no visible takers. (Economics 102 may tell them to move to Utah, where hiring is said to be surging. Tell that to a single mom, who had to borrow to pay for her courses.) In the summer of 2014, mighty Microsoft laid off 18,000 of its employees, saying their skills and services were no longer needed. Yet the company was concurrently lamenting that it was unable to fill several thousand key positions, due to a lack of qualified applicants. (I'll be returning to whether there's a contradiction here.)

But why the declining demand for electrical engineers? There was a time when firms like General Electric and General Dynamics, along with Westinghouse and Western Electric, were at the core of the workforce. Yet, unknowingly, these companies were sowing the seeds of a profession's decline. Using their own ingenuity and skills, their engineers were creating equipment and processes whose operators and implementers would no longer need the level of training they themselves had. Paul Beaudry and his colleagues at the University of British Columbia and York University call what is occurring "deskilling."

This trend itself has been a harsh surprise, since common wisdom has counseled that more of us would have to become smarter to cope with the complexities of our time. Yet the Beaudry group shows how "high-skilled workers have moved down the occupational ladder and have begun to perform jobs traditionally performed by lower-skilled workers." I'll be documenting how growing numbers of graduates with STEM degrees aren't working in those areas due to a paucity of jobs at what was supposed to be their levels. And this is happening despite their having accrued the intensive mathematics preparation they were told they would need.

Who is handling the tasks once the province of graduate

engineers? A 2013 study from the Brookings Institution found that "half of all STEM jobs are available to workers without a four-year college degree." Some of these workers have two-year diplomas, others earn certificates, but as many are high school graduates who honed new skills on the job. Here too, what's wanted is not traditional mathematics, but agility with numbers, usually geared to specific processes or equipment. Most such jobs are new, and more are being created every day. Here, courtesy of the BLS, I've listed just a few:

Gynecologic Sonographers
Avionic Equipment Mechanics
Stenocaptioners
Cryptanalysis Keyers
Logisticians
Electronic Drafters
Multi-Media Animators
Forest Fire Prevention Specialists
Continuous Mining Operators
Echocardiographers
Neuroscience Nurses
Synoptic Meterologists
Petroleum Pump System Gaugers
Geodetic Surveyors
Semiconductor Processors
Laboratory Phlebotomists
Environmental Inspectors
Tumor Registrars
Nuclear Monitor Technicians
Prosthodontists
Extruding Specialists
Digital Image Technicians
Electrophysiologists
Maxillofacial Radiologists
Pilates Equipment Designers
Remote Sensing Specialists

Engineering also offers a sad study in how the country mishandles the talent it has. As I've noted, the BLS estimates that in the decade ending in 2022, the United States will have created some 130,400 additional engineering positions. In 2013, the most recent year for figures at this writing, some 86,000 fledgling engineers graduated from American colleges and universities. At this rate, the decade will produce 860,000 new engineering graduates. But that number is over *six* times what the BLS is projecting for job openings. So let's look at what will happen to them.

In fact, the vast majority of those who do get engineering jobs will not be hired to fill newly created positions, but rather will be *replacing* individuals who are leaving the profession, either voluntarily or semivoluntarily. Businesses and other organizations that hire engineers find they can recruit the graduates they want by offering salaries acceptable to men and women in their twenties. What isn't said aloud is that the pay won't rise much higher.

This becomes evident from looking at *median* salaries for engineers, because the midpoint is about what a midcareer engineer makes. According to BLS salary surveys, in 2014, the medians ranged from $71,369 for civil engineers up to $96,980 for aeronautical engineers.* By way of comparison, the median for nurse practitioners works out to $83,980 and for pharmacists $101,920.

And, of course, a median means that half are earning less. This tells us that compared with other professions, engineers have a relatively brief employment span, making it largely an occupation for younger people. Those who start families tend to shift over to sales or middle management, so new graduates are hired, at under-median pay, to take their place. All this attrition could be solved by improving engineering salaries and creating career paths akin to law and medicine. But employers see no need to do that, since each year another 86,000 American graduates come on the market (not to mention holders of H-1B visas, whom I'll be discussing shortly).

The result is a huge waste of trained professionals, usually because their skills are never put to use. In a National Science Board

*Topping the chart at a special high of $130,280 are petroleum engineers. But this includes premium pay for desert sands and arctic ice, far from home and family life (as well as layoffs when the price of oil falls).

count released in 2014, fully 19.5 million American adults had scientific or engineering degrees, but only 5.4 million of them—28 percent—were working in STEM fields. A 2010–2012 survey of recent engineering graduates by the Center for Economic Policy and Research found that 28 percent of them were either unemployed or holding jobs not associated with their degrees. The figures were even worse for computer science and mathematics graduates, where 38 percent had failed to secure positions a degree holder should expect. (In fact, graduates in health and education were doing measurably better.)

THE SHORTAGE MYTH

Despite all these facts, the "shortage" drumbeat persists. "The United States is failing to produce enough science and engineering graduates," we hear from a business group called Achieve, Inc., which wants to deploy the Common Core as a remedy. Or this from John Cornyn, a U.S. senator from Texas: "As we all know, there is a scarcity of qualified people for many jobs, particularly in high technology." "We have a shortage," echoes Microsoft's chief counsel. "The shortage is going to get worse." And a panel advising President Barack Obama says annual output of STEM graduates must be raised by 30 percent "if the country is to retain its historical preeminence in science and technology." Why are they and many others talking this way? Are they unaware of the discouraging data on STEM employment we've just seen?

Let's take a look at what's happening on the ground. In 2014, a *New York Times* reporter visited a Wisconsin metal-fabricating firm, where the CEO complained that, of 1,051 recent applicants for openings at the firm, only ten of them could perform the technical operations he needed. The firm had the jobs, he said, but couldn't find qualified candidates. Further down, it was reported that the starting pay on offer was $10 an hour. What wasn't revealed was whether there were skilled people in the area who didn't apply because they wouldn't work for a wage very close to what they might get at McDonald's.

The BLS defines "shortage" as a "demand for workers in a particular occupation [that] is greater than the supply of workers who are qualified, available, and willing to do the job." So there isn't a shortage just because employers say there's one. We should also ask if potential employees are willing to take the offered jobs. They may have questionable reasons for not applying. But in a free labor market, their preferences are part of the equation.

Here it would be well to revisit Economics 103, which teaches that if a firm can't fill all the orders coming in, it will raise wages to attract the additional workers needed to get the shipments out. "If there's a skill shortage," Mark Price of the Keystone Research Center points out, "there has to be a rise in wages." But when today's employers talk of having unfilled positions, they're really saying they want it both ways. On the one hand, they want applicants who have the specified training and experience. On the other, they want those workers to accept ascetic pay. The Boston Consulting Group tells its clients that such talk is delusory: "Trying to hire high-skilled workers at rock-bottom rates is not a skills gap."

The Microsoft Corporation has been among the most vociferous champions of the shortage story. In 2014, Microsoft released a glossy thirty-two-page report titled *A National Talent Strategy*, which says our colleges are turning out only half the computer science graduates the economy needs. One of the report's suggestions is to start Microsoft-oriented education even earlier: "Broaden access to computer science in high school to ensure that all students have the opportunity to gain this foundational knowledge."

But among Microsoft's many statistics, a few don't appear. One is that the number of bachelor's degrees awarded in computer science decreased from 59,488 in 2004 to 50,962 in 2013—a 29 percent drop in 9 years, during a time when students were measuring college majors by their job prospects. Apparently, word got around that computer science graduates weren't being besieged with offers. Perhaps students also saw a survey conducted by the

American Association of Professional Coders. It found that half of its members were making under $41,000, with only eight percent earning as much as $70,000.

Another reason companies feel there is a shortage of qualified workers may have to do with changed expectations about suitable training. At a Minnesota company called Wyoming Machines, which creates and installs armor for military Humvees, the pay starts at $20 an hour, plus generous benefits. Yet, according to a *New York Times* report on the firm, the owner said she still couldn't find people with the kinds of skills she needed. We then learn that these Humvees need a novel kind of welding, which requires having a familiarity with the temperatures, gases, and pressures unique to this product. But few applicants come in the door with this specific combination of knowledge and skills.

Even if a nearby community college has courses in advanced welding, graduates are unlikely to arrive ready to armor Humvees. Alec Levenson at the University of Southern California points out that in most vocational programs, "skills are applied in a college context, not a workplace context." So calling for more STEM degrees won't achieve much if employers have very specific tasks they want new workers to perform on the morning they arrive. An apt case in point is this 2014 "Help Wanted" advertisement in the *New York Times:*

> Engineer III. Assist to process RFIs, Change Orders, Bulletins, Addenda. Develop specs, system narratives, and layouts. Ensure/Oversee QA/QC process. Establish sys application, design and op. parameters/sequences. Conducts field inspections. Coord w/other disciplines and work w/ design team. Creates designs. Work with CAD, AutoCAD, MS Suite, Adobe Acrobat, fluid flow, LEED principals, RevitMEP codes, and rel design and calc software incl. energy analysis sw.

This looks akin to dating-service participants, who prepare long lists of traits and tastes they require in any possible partner.

(Don't even bother if you don't share my passion for Thai food.) Companies used to take for granted that they would sponsor in-house training programs, which could extend for several months, if not longer. If they were costly, they were regarded as investments. Now firms want to do things as cheaply as they can; witness the $10 an hour offered to metalworkers and the $40,000 for degreed coders. At those rates, it won't be surprising if people with skills keep their eyes open for something better.

THE H-1B SOLUTION

Microsoft's call for *A National Talent Strategy* reiterated the company's claim that "many industries across the U.S. are unable to fill high-skilled American jobs with high-skilled American workers." As if to embarrass the nation where it is based, Microsoft pointed out that only four percent of U.S. college students graduate with engineering degrees, compared with around 30 percent in China.

Now, indulge me a paragraph that may not seem to belong in this chapter. Trust me, it does.

For at least half a century, America's agricultural industries have contended that they cannot find enough homegrown citizens to pick and pack crops. No matter how many "Help Wanted" placards were posted, even areas with high unemployment had few people signing up. Vacancies went unfilled—and crops rotted—due to grievous "labor shortages," or so the growers argued. Hence arose their claim that they had no option but to recruit from other countries. Today, almost all the domestic fruits and vegetables we consume are harvested by seasonal workers from Mexico, Central America, and the West Indies. As hardly needs saying, wages are low, work is seasonal, the jobs are literally backbreaking, with no benefits and few protections. Indeed, from this country's earliest days, employers have counted on desperation in other countries to keep their payroll costs down.

It's no longer just producers of grapes and broccoli who are asserting that their survival requires importing foreign labor.

Today we hear the technology sector making similar arguments. An outward difference is that this industry needs not hands and backs, but more cerebral skills. Hence its reliance on the widely debated H-1B statute, which grants special visas to certain classes of employees from other countries. As of the end of 2012, fully 262,569 holders of such visas were working in the United States. By far the most came from India (168,367), with China in a distant second place (19,850). Indians lead the list for an obvious reason: they come already knowing English. This is especially important for transient workers, who should be ready to carry out assignments the day they arrive. Among H-1B visas, "computer-related occupations" lead the list, followed by engineering. In fact, 2,619 were slotted for "occupations in art," which suggests an aesthetic side. But it turns out that many would be doing pictorials for video games.

Microsoft heads the list of American firms recruiting foreign workers. Intel and IBM follow, as do Hewlett Packard and Oracle. Indeed, a whole host of enterprises have software systems that need attending to. The H-1B roster also includes Rite-Aid, Goldman Sachs, Deloitte, and JP Morgan Chase.

The key question, of course, is, Why haven't more Americans been available for these 262,569 jobs? During the 2003–2013 decade, the most recent for which we have figures, our colleges graduated 460,726 majors in computer science, which seems a fair-sized pool. Yet, as we've seen, by no means all have found jobs associated with their degrees. Norman Matloff, himself a computer scientist at the University of California's Davis campus, has done the most extensive analysis of the H-1B pool. He has found that most are in their middle or late twenties, not far from the average for recent American graduates. Nor is there reason to think that computer science training in India and China is markedly superior to ours. So what makes them so attractive to employers?

The answer is very similar to the one that explains why our crops are harvested by workers from other countries. In both cases, non-Americans are willing to take jobs with terms of employment that are not attractive or viable by American standards. Typical young Indian programmers are single, or at least come unattached, and are willing to commit to a five-year stint to jump-start their prospects back home. They don't expect to be promoted, so their employers don't have to worry about creating career paths. And then there's the matter of pay. Zoe Lofgren, who represents much of Silicon Valley in Congress, released figures for 2011 which showed that H-1B workers averaged only 57 percent of what was received by Americans with comparable credentials. Visa applicants must be "sponsored" by a particular employer. Once they are here, they cannot leave that employer, either for another offer or if working conditions turn sour. Ross Eisenbrey of the Economic Policy Institute sums up: "they are more or less indentured, tied to their job, and whatever wage the employer decides to give them."

Regardless of whatever "skills" they bring, few expect or are given high-level assignments. Most of those admitted under H-1B are, as Matloff puts it, "ordinary people doing ordinary work." Typically, they sit in cubicles, turning out the endless lines of code needed to keep much of the world humming. Of course, what they're doing seems mysterious to most outsiders. (In another chapter, I'll offer a glimpse of what codes actually looks like.) Yet according to the General Accounting Office, less than 17 percent of the H-1B pool comes classed as "fully competent."

Coding is writing instructions, in a mode a machine can follow. Billions of lines must be contrived. Those supervising projects can find the work challenging. But beneath and behind each creative designer will be dozens, even hundreds, of coders who must get every symbol, letter, and integer precisely right. One erroneous keystroke can jeopardize hundreds of thousands of enrollments in a national health program. The STEM explosion of our time is built on hunched backs and bleary eyes in hundreds of thousands of backroom cubicles.

The "shortage" fiction, in tandem with the H-1B strategy, offers insights into a corporate plan for the coming century. We've always known businesses like an oversupply of workers, in part to keep those with jobs fearing that they'll be replaced. And, of course, the less money that has to go to the rank-and-file, the more there will be to distribute among the executive suites. Matloff writes that the "shortage" mantra is actually "all about an industry wanting to lower wages."

Sharper income differentials and new class configurations coalesce in this corporate vision. But unlike in earlier eras, in our time those at or near the bottom will be expected to have a level of technical literacy. As we've just seen, for each analyst who conceives the next breakthrough, hundreds of coders are needed to get the keystrokes exactly right. (Recall the Cratchits and Bartlebys who sat entering sums in ledgers.) They will be a high-tech proletariat. As was noted earlier, only one in ten will reach even $70,000. What they'll do as they reach a hoary age like thirty-five is not the concern of an economy based on high skills and low wages.

TOYOTA CHOOSES MISSISSIPPI

A panel on mathematics education, convened in 2008 by Margaret Spellings, then George W. Bush's secretary of education, announced that nearly half of young Americans were unable to "do mathematics at the level needed to get a job at a modern automobile plant." This was a shattering indictment. After all, over two-thirds of recent high school graduates have now had at least two years of algebra. If they can factor trinomials, why are they so lacking in workplace skills? Or perhaps that's the wrong question. Might it be that the panel, which was composed almost wholly of academics, leveled the charge without having actually visited a factory? Lynn Arthur Steen of St. Olaf's College has an answer: "mathematics teachers simply do not know much about how mathematics is used by people other than mathematicians."

I decided to do a reality check of my own. Some of this country's most sophisticated manufacturing takes place at factories under German and Japanese ownership. So I looked at where BMW, Nissan, and Honda have located some of their plants. Here are four locations they've chosen:

Where High-Tech Carmakers Locate

Company	Location	Dropout Rate*
Nissan	Coffee County, Tennessee	134%
BMW	Spartanburg County, South Carolina	137%
Honda	St. Clair County, Alabama	146%
Toyota	Union County, Mississippi	161%

* Compared with nationwide dropout rate (= 100%)

Spartanburg County in South Carolina, Tennessee's Coffee County, St. Clair County in Alabama, and Union County in Mississippi. I'll concede they select regions that are not hospitable to unions. But it seems safe to surmise they also want workers fully able to deploy their state-of-the-art equipment. Today's vehicles are guided by computer chips and equipped with all manner of electronic bells and whistles. Automotive assembly today calls for a lot more than installing seats and windows. Yet, what I also found notable about these counties is that their school systems are far from impressive, having dropout rates palpably above the national average. Even so, all of these firms turn out very reliable vehicles. And they do this with local workers, almost all of whom were raised in the area, who are *less likely* to have studied algebra than their agemates elsewhere. These employers feel that if applicants have good work habits and attitudes, they can be taught what they need to know.

Well, what about mathematics? Linda Rosen, an economist at

the National Alliance of Business, found that "most problems in the workplace involve applications of basic arithmetic: addition, subtraction, multiplication and division." Still, I made some inquiries myself. I learned that Toyota, the Mississippi company, collaborates with nearby Northeast Mississippi Community College. The college willingly devised a *Machine Tool Mathematics* sequence geared to the skills assembly workers need. Mike Snowden, who runs it, told me they avoid conventional textbooks, so as to focus on "algebraic and trigonometric operations essential for machining."

In fact, formal schooling in algebra can be counterproductive. John P. Smith of Michigan State University has had a long career of research in real-time factories. "Mathematical reasoning in workplaces differs markedly from school mathematics," he informed me. "In fact, the algorithms taught in school are often not the computational methods of choice for workers." Not the least reason for this, he added, is that "very few teachers have any idea about what goes on in the work world."

WHO CAN MEASURE A RECTANGLE?

With all the emphasis on mathematics, the importance of arithmetic is too often overlooked. As Steen notes: "What current and prospective employees lack is not calculus or college algebra, but a plethora of more basic quantitative skills that could be taught in high school, but are not." In fact, arithmetic-based quantitative skills can be just as demanding as differential equations. But mathematics mandarins view this as dumbing down and oppose allowing any alternatives to their syllabus.

Remember the corporate-sponsored American Diploma Project, which told us that even upholsterers will be needing advanced mathematics? To make their case, they cited a test from the Federal Aviation Administration, designed to ensure that competent mechanics will keep our aircraft safe. Here's one question they reproduced. As it happens, I like it, largely because it involves no advanced mathematics at all. Indeed, I gave it to my college class in numeracy, all of whom had had geometry, trigonometry, and two years of algebra. Sadly, not all of them answered it correctly. Try it yourself.

FAA Airframe and Powerplant Certification Assessment

A rectangular-shaped fuel tank measures 27½ inches in length, ¾ of a foot in width, and 8¼ inches in depth. How many gallons will the tank contain?

(231 cubic inches = 1 gallon)

(a) 7.366 gallons (b) 8.839 gallons (c) 170,156 gallons

To get the right answer, you have to note that while the tank's length and depth are gauged in *inches,* its width is given as a fraction of a *foot.* This is certainly not the usual way an object's dimensions are recorded. (It's almost on a par with giving one of the dimensions in centimeters.) Too speedy a scanning of the question caused students to overlook that mismatch. Without thinking twice, they might simply multiply ¾ (or .75) by 27.5 inches and then by 8.25 inches to obtain the tank's cubic volume. If they do, they'll get the erroneous 7.366 gallons, which the test makers obligingly provided as an option. Rather, the ¾ should be transposed into 9 inches, and only then does the three-way multiplication yield the correct answer (b).

I especially like this test, insofar as one of its aims is to ensure that aircraft mechanics read all instructions extremely carefully and take a slower look if some numbers seem suspicious. With our penchant for fast-paced examinations, we are undervaluing those attributes and may well pay a price for it. After all, airplane mechanics, no less than pilots, have our lives in their hands.

VOICES

I live in Silicon Valley and know professionals in all manner of technical fields. Yet I can only think of one person, an aeronautical engineer, who uses advanced math to actually do his job.

In other nations, such as Germany and Switzerland, it would be considered absurd to say that all sixteen-year-olds should have to learn the same stuff.

Algebra is probably one of the main reasons for the decline of American business. The tinkerers and inventors who were this nation's industrial titans could not even get into business schools today, with their mind-numbing emphasis on math.

As a lawyer, I use addition, subtraction, multiplication, division, and percentages. But algebra, trigonometry, calculus, geometry? For 99.9 percent of use, they were wasted time and effort.

VOICES

I have a CPA certificate. To this day, I have no idea what trigonometry is useful for. Another example is the business school requirement of calculus. Never used again.

I am a retired finance type of guy with an MBA. In my forty years of work, I have never had to solve or use quadratic equations. The times tables and long division usually sufficed.

I was a horrible algebra student. The only benefit from having to take the course was getting this cute guy to tutor me. I still failed, but this year he and I will celebrate our forty-fifth anniversary.

As a student, I only squeaked by in math because my parents could afford a coach. I have had a successful career in marketing, where only arithmetic is needed.

4

Does Your Dermatologist Use Calculus?

A study by the Association of American Medical Colleges asked 14,240 medical students to rate how various premedical courses related to their clinical studies. Predictably, 82 percent said biology was important, 79 percent agreed about biochemistry, and 65 percent put comparative anatomy on the list. At the bottom of the list was calculus, with only 3 percent, and most of those who cited it were aiming at careers in research. Still, many medical schools—including Johns Hopkins, Harvard, Duke, and all those in the California and Texas systems—insist that applicants take a full menu of advanced mathematics including calculus. As one physician wrote in response to my *Times* article: "For medical school, one big hurdle was always calculus, a thoroughly irrelevant course. Any honest physician will tell you the last time he/she used calculus was on a final exam in the subject." Or this from Penny Noyce, who combines her practice of internal medicine with writing children's books: "For entry to medical school, I was required to take calculus; but I never needed it for learning anatomy or pharmacology, and I certainly never used it in patient care. Calculus was a rite of passage, perhaps a weeding-out process."

As is well known, there are many more candidates than places, and examining all those files takes huge amounts of time. One way to deter people from applying is to pile on even more premedical requirements, with calculus a case in point. (All the more, because medical applicants are expected to get top grades in all their subjects.) The end result is to bar many men and women who would turn out to be superb physicians. Due to a pointless barrier, they have no chance to display their talents.

Richard Kayne, a physician and admissions dean at New York's Mount Sinai Medical School, assured me that "only arithmetic is needed in patient care." He agreed that practitioners should be sufficiently at home with numbers to understand the statistics used in journal articles and conference papers, such as knowing whether a correlation described as $p < .01$ is stronger or weaker than one designated $p < .05$. I then showed him a question on a Medical College Admissions Test:

Two charges (+q and −q) each with mass 9.11×10^{31} kg. are placed 0.5 m. apart and the gravitational force (F_g) and electric force (F_e) are measured. If the ratio F_g/F_e is 1.12 x 10−77, what is the new ratio if the distance between the charges is halved?

(Time to read and answer: 80 seconds.)

(a) 2.24×10^{-77} (b) 1.12×10^{-77} (c) 5.6×10^{-78} (d) 2.8×10^{-78}

Correct answer: (b)

Dean Kayne confessed he couldn't solve the problem, nor could he find any rationale for its use in medical admissions. True, there are only a few mathematics questions on the MCAT. Still, they count for precious points in a competition for scarce places, despite the fact that they reveal nothing about what it takes to be an effective physician.

Anthony Carnevale and Donna Desrochers, who analyze the kinds of skills workers need, worry that irrelevant job requirements are denying entry to talented individuals who can become top performers. "Higher levels of abstract mathematics are required for access to certain professions," they write, "even when high-level mathematical procedures are unnecessary in the day-to-day work of those professions."

ACTUARIES

Actuaries provide a case in point. Of occupations dealing with numbers, they surely rank high on the list. It's their job to compute the likelihood of a hurricane hitting a coastal county, how many infants will contract a new variant of influenza, or how much to raise your car insurance if you report a second dent. And of course mathematics is vital to computing these and other probabilities, especially algebra, when many interacting factors figure in the equations. So the question is not whether mathematics is necessary—it is—but how much.

To become an actuary, one must pass stringent examinations set by senior members of the profession. After a careful look at them, I've concluded that they aren't far from the final round for a Princeton PhD in mathematics. Candidates are expected to master Gaussian distributions and Markov chains, along with Brownian motion and Chapman-Kolmogorov equations.

I mentioned this to Gabe Frankl-Kahn, an actuary at TIAA-CREF, a multibillion dollar fund, which superintends pensions for more than a million college professors. While Frankl-Kahn passed the test early in his career, he confessed that the "test covers mathematics that people will never need in their jobs." He said he doubted if he could pass it again if asked to do so today.

Markov and Kolmogorov aren't on the tests to ensure actuarial competence. Rather, they're included as part of an effort to raise the profession's stature. In an age preoccupied by status, mathematics serves as a surrogate for precision and rigor. As more requirements are piled on, the more prestige will supposedly accrue to an occupation. Even educated people are awed by Kolmogorov, while they haven't the slightest idea who he was or what he did.

COMPUTING CAREERS

Of all occupations with a foothold in the future, those attuned to computers lead the pack. Whether software developers, systems analysts, or mainframe engineers, those working in these fields are seen as the forefront of a digital century. For my own part, I wholly support the knowledge and skills associated with their hardware and software. (I've used the web daily while writing this book.) At this point, all I want to show is that mathematics figures far less than one would think in this realm.

David Edwards, a mathematician at the University of Georgia, teaches a course called Mathematics for Computer Science. But he often wonders why it's being offered. He told me of a visit by a team of recruiters for a firm that develops software for clients. He invited them to meet with his students, since they said they were looking to hire mathematics majors. Edwards continued:

> After they finished, I asked them, "What mathematics do you actually use?" They sheepishly responded, "None." So I asked, "Why do you specify mathematics in your hiring?" They told me it was a "convenient filter," to whittle down the number of applicants they'd have to interview.

So if not mathematics, what is needed in computing occupations? I walked across my own campus, to observe some computer science courses. The ones I chose, with the professors' permission, were Software Engineering, given by Sy Bon, and Alex Ryba's Algorithmic Problem Solving. I was entranced by both classes. Indeed, the time flew by, even if I grasped only a fraction of what the instructors or students said. Even so, I sensed that what was being taught was integral to our lives, a body of knowledge and understanding on which our brave new world depends. That said, I will now add that in the one hundred fifty minutes the classes were in session, not a single mathematical notation or equation was alluded to, or projected on a screen or board. While computer programs use numbers as raw materials, the codes that do the organizing are almost entirely composed of symbols.

For a glimpse of what computer scientists are expected to know, I've reproduced a question from the College Board's Major Field Test in the subject, which is taken by college seniors applying to graduate programs. It's all those *A*'s and *j*'s and *k*'s that store data in your laptop and let you memorialize your pet cat on YouTube.

If *A* is an array with *n* elements and procedure *Swap* exchanges its arguments, then the following code segment sorts *A* in descending order:

```
For ( int j = 0; j < n - 1 ; j++ )
For ( int k = 0; k < n - j - 1; k++ )
If ( A[k] < A[k + 1] )
Swap( A[k] , A[k + 1] )
```

How many calls to Swap are made initially, A[I] = 1, for I = 0, 1, 2, . . . n - 1?

(a) $n - 1$ (b) n (c) $n(n - 1) / 2$ (d) $(n - 1)(n - 2)$ (e) $n(n - 1)$

Correct answer: (c)

In fact, there's a lot of myth about how much mathematics is needed to create all the technological wonders seemingly invented almost hourly. Developing a single video game may call on the talents of several hundred people. So a costume designer may turn to a mathematically versed colleague and say that she wants the villain's coat to snap like a whip. He'll know what equations to use. But she doesn't have to. Nor apparently does a software designer, who told a survey conducted by MIT, "I don't really use much of any math, except for calculating the tip at lunch."

ENGINEERS

Engineering covers a spectrum of specialties, so it's best not to generalize. In *UME Trends*, Robert Pearson writes: "My work has brought me into contact with thousands of engineers. But I cannot recall more than three out of ten who were versed enough in differential equations to use them in their daily work." Sunil Kumar, who heads mechanical and aerospace engineering at New York University's Polytechnic Institute, told me that 90 percent of the tasks in his two fields no longer call for mathematics. At Consolidated Edison, the gas and electric utility for most of metropolitan New York, I met with Paul Miroulis, Peter Van Olinda, and Kenneth Chu. After putting their heads together, they judged that of the company's six hundred electrical and civil engineers, probably no more than eighty used mathematics in their jobs.

Julie Gainsburg trains high school mathematics teachers in the school of education on the Northridge campus of California State University. Unlike many of her colleagues, she believes teachers should know how their subjects are used in the outside world. To this end, she spent several weeks observing engineers who were supervising construction of a nearby apartment complex. At one point, she peered over their shoulders, as they were confirming that some trusses would bear a building's weight. What struck her as an educator was how "the emphasis for engineers differs dramatically from school courses." She adds: "Algebra usually offers a heavy dose of complicated algebraic manipulation. In contrast, the manipulation I observed in structural engineering work was always simple, involving only one or a few basic operations."

At most, virtually all they used was multiplication and division. Indeed, "only once did I see an engineer representing a situation with an original algebraic formulation." And that was essentially a middle-school exercise, representing the cross-section area (*A*) and the diameter (*d*) of a metal rod. Here's all she saw on his pad: $A = \pi d^2/4$. This accords with what the University of Georgia's David Edwards has seen. The vast majority of engineers, he says, use "only eighth-grade mathematics."

Of course, engineers should be taught the kinds of mathematics they'll need when they practice their profession. But by and large this isn't happening. Most engineering programs hand mathematics instruction for their students over to their university's mathematics department. On paper, this may look sensible. After all, they know the subject, while engineering faculties have their own specialties and don't want to be saddled with teaching trigonometry. So what the students get are further doses of *academic* mathematics: more of the abstract exercises they've been given since the eighth grade. Only rarely do mathematics instructors, whether in middle school or at the college level, have any awareness of how aspects of their subject are used by engineers. Or, for that matter, by actuaries or software developers. (How many mathematics professors have ever invited a practicing chemist to their classroom?)

I asked Mitzi Montoya, while she was a dean of engineering at Arizona State University, whether her students really need to take calculus and algebra. Her response was straightforward. "If you go out and look at what engineers use, it's not calculus or differential equations," she told me. "Even if you go into a big company that's building sophisticated rockets, you would still only find a very small percentage that are doing mathematical analysis." She went on to say that engineering is highly quantitative, and her students needed to know how numbers function in the real world. Sadly, they didn't get that in the academic mathematics courses they had to take. "So one thing we designed," she told us, "was our own mathematics course which was very problem based." Not surprisingly, this replacement threatened some mathematics professors on the campus. They used their power to shut it down.

Dean Montoya is also troubled by how engineering schools pile on mathematics requirements for admission. She cited some applicants to Arizona State who had completed exciting robotic projects in high school, but couldn't cut it with calculus. It was an uphill fight to get them into her program. To her mind, we may be losing incipient Edisons to succeed Steve Jobs and Bill Gates.

SCIENTISTS

"Many of the most successful scientists in the world today are mathematically no more than semiliterate," wrote E. O. Wilson, one of our era's most prominent biologists. (If there were a Nobel Prize in his field, he would have received it long ago.) Wilson reminds us that Charles Darwin had no talent for mathematics, nor was it the basis for his theory of natural selection. Even as the scientific world has grown larger and more complex, he notes that "mathematical fluency is required in only a few disciplines." He cites information theory, particle physics, and astronomy, but adds that fields like these make up only a small corner of the scientific enterprise.

Carl Friedrich Gauss called mathematics "the queen of the sciences." That's not surprising, since he was a mathematician. Even so, in recent times, it's not easy to cite scientific breakthroughs derived from the work mathematicians do. Tony Chan, a UCLA mathematician, points out that "mathematics generally has not been seen by other scientists to be at the frontiers of science." Avi Loeb, who chairs the astrophysics department at Harvard, is even less impressed with what mathematicians do. "In physics, you are required to base what you do on proven facts," he points out. "In mathematics, you are allowed to go in all directions that have no connections with reality." Of course, numbers figure strongly in science; peruse any journal article. But most of the tables are based on arithmetic, garlanded with tests of statistical significance.

What also worries E. O. Wilson is that mathematical hurdles as early as high school have "deprived science of an immeasurable amount of sorely needed talent." So along with thwarted physicians and engineers, we must now add a host of stillborn scientists.

John Matsui, a biologist on the University of California's Berkeley campus, agrees that some mathematics is needed in his field. But not much. "We have a new mathematics course," he told me, "to replace the general calculus course once required for biology majors." In fact, very few if any biologists will need or use calculus. But when they approach "mathematics as problem solving," he said, "its skills are transferable to a field like biology." The Calculus 201 in a mathematics department doesn't help budding scientists. They need a mathematics tailored to their discipline, which few mathematics faculties are willing, let alone able, to teach.

Carl Wieman, a Nobel Prize–winning physicist at Stanford University, has given as much attention to teaching as to research. He sees the emergence of two camps of scientists: "theorists" and "experimentalists." But today's theorists are not much like the solo savants of the past. Rather, they rely on computer-driven models, which churn out equations that try to replicate reality. Wieman himself is devoted to actual experimentation. "The trick," he told me, "is to formulate simple mathematical approximations of more complex functions." Since the real world is too blurry for ultraprecise solutions, he added, researchers "use more sophisticated mathematics less and less."

SANS MATHEMATICS

"For more than forty years, I have been speaking prose without knowing it," remarked Monsieur Jourdain, in Molière's *Le Bourgeois Gentilhomme*. Each day, all of us use mathematics to make decisions and formulate ideas. Nor am I talking here about plain vanilla arithmetic, as when we verify our bank balance or compute a restaurant tip. The mathematics we utilize can be quite subtle and complex, even if we do not realize it.

I have on my desk a heavily footnoted research study entitled "Mathematics Practice in Carpet Laying." It is fully empirical, based on close observations of workers on the job. The author begins by noting that, at best, the installers had finished only high school. None had been stellar students, nor could any recall lessons from geometry classes. Yet each day they put into practice precepts reflecting that branch of mathematics.

Laying carpet has two principal rules. The first: in cutting pieces from the large roll, leave as little unused material as possible. The second: the finished layout should also show as few seams as can be managed. Leftover pieces are viewed as unprofessional, while clients want a close-to-pristine surface. Carpeting around pillars, dormer windows, and spiral staircases presents special challenges.

The researcher spent several weeks closely watching a carpet team complete several jobs. In her later analysis, she said that she had observed them using *coordinate geometry*, including *points of tangency* and *computational algorithms*. Needless to say, these weren't terms used by the installers, who were unaware that they had been employing tangencies and algorithms. All had learned what they knew and did on the job, by assisting, watching, and asking questions. That they developed this quite intricate competence suggests that most of us have inherent mathematical aptitudes, which surface less in classrooms than in real jobs. Indeed, there is little reason to believe they would have done better at carpet laying had they studied more mathematics in high school or college.

Every racetrack has a circle of handicappers, who earn their livings selling tips to bettors who attend. (Recall the character called Nicely-Nicely in *Guys and Dolls*, who sang, "I've got a horse right here, his name is Paul Revere.") Their record for picking winners is well above average, so much so that they are viewed with respect, and often with awe. Nor is their success due entirely to instinct or intuition. The handicappers work with mountains of data, their principal sources being dense lists of numbers on the performances of all the horses running each day. These statistics cover dozens of variables, ranging from horses' lifetime winnings and quarter-mile times, to weather conditions during their races and penchants for hugging the rail. Handicappers synthesize these and many other factors, assessing how they interact and what weight to give to each.

In preparing a study called "A Day at the Races," a team of quantitative psychologists observed fourteen handicappers who frequented a Delaware track. As with the carpet study, each was first asked about his formal education. It turned out that nine of the fourteen had left high school before graduating, and none had taken a class in algebra. Yet despite this lack, they used the equivalent of algebraic equations recurrently each day. Twelve of the fourteen agreed to take an IQ-type test that contained mathematical problems. Most scored below the national norm.

The researchers were intrigued by the handicappers' success in picking winners. So they tried to translate their choices into mathematical terms. Thus one table they created was titled *Standardized Regression Coefficients Showing Net Effect of Interactive Model Variables on Log-Odds for Each Subject*. Another chart contained over a hundred regressions (from +0.75 to -0.46), correlating figures in the horses' racing records. The scholars concluded that while the handicappers were unusually skilled at permuting and combining thousands of variables, they had a type of mathematical intelligence that couldn't be assessed by standardized tests. The reason: most academic measures are "unrelated to real-world forms of cognitive complexity."

MATHEMATICS AND MAKING MONEY

Harvard's Tony Wagner has studied what businesses want in and from their employees. Even at high-tech firms, he reports, "knowledge of mathematics did not make the top-ten list of the skills employers deem most important." His finding also confirms that a need for mathematics has been highly overstated. And yet we continue to hear that studying more mathematics raises your chances of achieving higher earnings. Like most myths, this one sounds plausible at first hearing. After all, increasing your stock of skills should enhance your value where you work. Thus the business group called Achieve, Inc. has publicized a study purportedly showing that "the annual earnings of students who had

taken calculus in high school were about 65 percent higher than the earnings of students who had only completed basic math."

It is undoubtedly true that calculus completers tend to end up with visibly higher earnings than their basic-math classmates. But that correlation doesn't necessarily tell us that one event or condition has caused the other.

At last count, only 17 percent of all high school seniors had finished calculus, the most advanced course in the customary mathematics sequence. It is most usually offered in better-endowed schools, which are sought out and supported by parents who are better off. A full mathematics program is part of the grooming process for selective colleges and the careers they are supposed to ensure. But it isn't having studied calculus that brings the 65 percent income premium. It is early entry on a high-status escalator. In those same high schools, they may also have been made to study *Great Expectations*. Should we argue that reading Dickens leads to higher earnings? I might like to think so, but in honesty I can't. What counts is having been put on a college-prep track, regardless of what you studied.

When *A* and *B* are observed together, it's often wise to look for a *C* that may have caused *both A* and *B*. Yet this simple fact is seldom taught in mathematics classes. (Probability may be included, but causality typically isn't.) If young people were allowed to spend more time with commonsense statistics, they could make better sense of the world, regardless of how much they end up earning.

VOICES

Many of the peers I work with are very good engineers, but have retained very little mathematics. The reason they have retained so little is because they do not need to use it.

When I meet some of our students after they have graduated and taken engineering jobs, I like to ask them what mathematics they use in their work. The most frequent response: addition, subtraction, multiplication, division.

I have been an engineer for all of my adult life. Algebra? Calculus? Differential equations? I have forgotten most of all this from lack of use.

I have worked in a technical capacity at Texas Instruments and Honeywell and have been awarded two patents. I have never had to solve a calculus problem or a quadratic equation.

VOICES

It was algebra that made me drop out and join the U.S. Marines. I went to artillery school as a fire-direction controller. Pinpoint accuracy was our stock in trade. Since then I have spoken at high school classes about my experience. I start by telling them it was because of failing algebra that I joined the Marines.

I've taught in Japanese higher education for thirty years. No program at our university except engineering has a mathematics component.

The chief engineer on the Mars Rovers said he failed geometry twice. He adds that his teacher gave him an F+ so he didn't have to see his face again.

5

Gender Gaps

Whether or not there is a "mathematics gene," there's no evidence that it is stronger in one gender rather than the other. Here, I have to concur with Christiane Nuesslein-Volhard, who won the Nobel Prize for discovering how genes actually work:

> In mathematics, there is no difference among the intelligence of men and women. There is certainly no difference in the genes. The difference in genes between men and women is simply the Y chromosome, and the Y chromosome has nothing to do with intelligence. And all the other genes are present both in men and in women.

To be sure, mathematics has long been dominated by men, as have many other professional fields. Until 2014, all fifty-two recipients of the Field Medal had been men, as all Abel Prize winners still are. But black actors couldn't win Oscars until they were cast in eligible roles. What's been the case historically does not establish eternal verities.

As it happens, debates about gender-based capacities don't arise in other intellectual areas. We haven't heard it argued that women are less equipped for literature or history or anthropology.

One reason may be that mathematics maintains an abundance of data on grades and scores. I'll be drawing on some of these statistics in this chapter.

IN ACTUAL CLASSROOMS

Girls and women perform more ably in mathematics when they are assessed by the instructors who have come to know their work during a semester. A study by the National Center for Education Statistics scanned the transcripts of over 25,000 high school seniors. They found that the girls averaged 2.76 in the mathematics courses they had taken, against 2.56 for the boys. Another survey, this time of some 50,000 students in college mathematics courses, reported that 24 percent of the women got As, slightly ahead of 22 percent of the men. Granted, both margins aren't huge. But they tell us that teachers and professors felt the girls and women were absorbing more.

That the girls do better is noteworthy, not least because male classmates often dominate the lessons. An observational report, in the *Psychological Bulletin*, found that "boys are more active in providing answers, particularly unsolicited answers, while girls typically won't shoot up their hands first." The study noted that even women teachers called more on boys—all that arm waving isn't easy to ignore—despite suspicions that some of the girls had a firmer grasp. ("Jason, would you put your hand down? I'd like to hear from Jennifer.") Indeed, the answers the boys volunteered were often wrong, suggesting they hadn't fully attended to the questions.

Another study asked samples of high school students how they felt they were doing in their classes. The boys' self-ratings averaged 82, while the girls gave themselves a discernibly lower 64. By contrast, when the teachers rated the same students, their average grade for the boys was 76, while they gave the girls an 80. That the boys overrated themselves might be dismissed as male bravado. But that the girls undervalued themselves by sixteen points should be a cause for concern.

All the more, since the study found that "girls more than boys were less confident that their solutions were correct," even when they were likely to have them right. Teachers are well aware of such self-deprecation, not least because it usually isn't warranted. Indeed, it was lamented in a joint letter by the presidents of Princeton, Stanford, and MIT, who told the *Boston Globe* that "low expectations can be as destructive as overt discrimination." (This was a not-so-veiled retort to remarks by a fellow college head.)

Leonard Sax, a research pediatrician, found that "if the environment is right, girls can excel to the same degree and in the same subjects that boys do." What Sax says is affirmed by the record of women-only schools, which teach a full range of subjects, including exacting mathematics. Indeed, they have long been laboratories revealing the full range of women's abilities. At colleges such as Wellesley and Bryn Mawr, higher proportions go on to earn mathematics doctorates than do women from coeducational institutions.

ALGORITHMS AWARDING GRADES

A disturbing trend of our time is an almost reflexive disparaging of teachers. I find it ironic that such mistrust is seldom voiced by parents or pupils. Rather, the podium has been captured by critics, many of whom have not been near a classroom for years. Among the loudest have been state legislators and corporate executives.

One result has been efforts to minimize instructors' evaluations and give center stage to so-called objective tests, intimating that these instruments are shielded from subjective bias. With these tests, all students face identical or basically similar questions, and the software will show whether they have given the responses the testers were looking for. In most formats, pupils are told to pencil in bubbles on a sheet. In some, they write numbers in small squares. In both cases, computers scan the sheets and tally up the scores. (In many districts, pupils are already sitting at keyboards, with their answers submitted electronically.)

If teachers' judgments suffer from subjectivity, uniform tests eliminate the human element, supplanting classroom grades with spreadsheet statistics. This, at least, is the standardized story. Now let's see how it plays out on the gender front.

While women are getting better classroom grades, machine-graded tests tell a different tale. The 2013 National Assessment of Educational Progress test of high school seniors found girls averaging 304 in mathematics, behind the 308 recorded by boys. A test used in an Educational Longitudinal Study scored girls at 50.2 and boys at 52.2. True, these margins aren't huge. But bear in mind, these very same girls had outperformed boys in their classrooms.

The American College Test, reviewing the scores in its files, declared that 47 percent of the boys who took its test in 2013 were ready for college mathematics, while only 41 percent of the girls were. And on the Scholastic Assessment Test, girls averaged 499 in mathematics, discernibly behind the 531 for the boys. Gaps like these have been in evidence since the ACT and SAT began. More striking, almost twice as many of the boys (9.6 percent) had SAT scores in the top 700–800 tier as did girls (5.2 percent). Hence, a two-part question. The first is whether the standardized gaps show that women are in fact not as accomplished as their teachers say. The second is how far the multiple-choice format allows most people, but especially women, to show what they know and can do.

It has long been accepted that the Godzilla of standardized testing is a rapidly ticking clock. Typically, the SAT allocates an hour and a quarter to cope with sixty mathematics problems, which comes to 75 seconds for each one. "Boys' brains are built for speed," Leonard Sax, author of *Why Gender Matters*, was quoted by the *Wall Street Journal* as saying, whereas "girls' brains are built for complexity." In my own classes, when I ask young women a question, they are more apt to reflect for a few moments before hazarding an answer.

Given their affinity for speed, boys start by sizing up multiple-choice questions with a quick glance. The Educational Testing Service, which runs the SAT, found that boys save time by mental strategies "that enable them to see the solution without actually working out the problem." As a result, they show less hesitation about racing ahead. And this usually pays off, since the SAT format doesn't ask or care how you got to your answer. Another tactic that comes more readily to boys is to eliminate, say, three of the options as plainly wrong, and then guess between the remaining two. This willingness to gamble shaves the odds in their favor. Another ETS study was even more revealing. It found that because girls spend more time pondering, they are measurably more likely than boys to leave some questions blank, and hence fail to reach the end. In all instances, girls' greater penchant for reflection undercuts their scores.

Taken together, a proclivity for speed and a readiness to guess are rewarded in clock-driven tests. And it's traits like these, not knowledge or skill in mathematics, that raise boys' scores. I hope no one will argue that quickness attests to depth of understanding. I recall a *New Yorker* article on Grigori Perelman, often identified as the foremost mathematician of our time: "He always checked very, very carefully. He was not fast. Math doesn't depend on speed." How would he have fared on a test that allowed only 75 seconds per question?

The study, which I cited earlier, showing that women were more likely to get A's in college mathematics classes also analyzed the SAT scores of both genders. Interestingly, the researchers found that the men who were awarded A's had averaged 623 on the SAT, while the women receiving the same grade got only 574 on the same test. Yet the ACT and the SAT both claim that they can forecast how students will fare once they are taking college courses. That, presumably, is why admissions officers give those tests such close attention. They also point out that grading may vary considerably among schools. So two applicants can have similar transcripts, but differ deeply in academic skills. Scores on a common test enhance admissions evaluations. Yet despite these and other arguments, there is no shortage of studies showing that standardized formats are egregiously inaccurate in predicting what women can and will achieve in mathematics.

HOW TESTS TWIST THE GENDER BALANCE

With women getting better grades, it shouldn't be surprising that more of them are attending and completing college. Of the bachelor's degrees currently being awarded, fully 57 percent go to women. Put another way, for every one hundred women in the nation's commencement processions, there are only seventy-five men. So it is even more dismaying that while more women are succeeding in education, they continue to be denied awards that are supposedly based on merit. In particular, they are not receiving what should be their share of places in top-tier schools.

As I showed in an earlier chapter, highly selective colleges expect scores above 700 on *both* the verbal and mathematics sections of the SAT. (To be sure, lower bars are set for alumni offspring and athletes.) But because almost twice as many men get 700 on the mathematics segment, more of them are propelled to the top of the acceptance list. This imbalance allows elite schools to maintain at least a 50:50 gender ratio, despite the 57:43 women-to-men ratio in the rest of the country. While Harvard and Columbia will never admit it, they don't want more women than men. That would affect the image and ambiance they wish to preserve, not to mention male alumni largesse. That's why, in 2014, they capped their women's enrollments at 48 percent, while Stanford chose 47 percent.

For the record, I should add that the reverse is true at MIT. On paper, it has always been open to admitting women. Still, in 1964, there were only ninety-eight women on the campus, or three percent of the student body. Two decades later, in 1984, the share of women had risen to 19 percent. More recently, MIT has decided to take coeducation seriously. (Among other things, both its president and admissions director have been women.) In 2014, women accounted for 45 percent of all undergraduates, quite close to Columbia's 48 percent and Stanford's 47 percent.

As in the past, MIT continues to expect astronomical mathematics scores from the men it accepts. But given the paucity of women who score 750 and above on the mathematics portion, applying a strict cutoff would render MIT only barely coed. Hence I found myself musing that its 45 percent could occur only if women were cut some slack on their mathematics scores. So I asked MIT for the score ranges by gender among the students it accepts. The admissions office replied that it couldn't tell me, because it doesn't assemble such numbers.

ONE CITY'S SELECTIVE SCHOOLS

New York City has long prided itself on its specialized public high schools, where the quality of instruction often rivals private academies. For almost all of these schools, entry is based solely on a three-hour test, taken by students at the end of the eighth grade. (The exceptions are schools that focus on music and performing and fine arts.) The competition is fierce. In 2013, Stuyvesant high school had 22,675 applicants, and sent 963 acceptances. Brooklyn Latin had a similar ratio: 14,147 candidates for 526 spots.

So here's another set of statistics. In the citywide pool taking the test, 51 percent were girls. Yet girls were offered only 45 percent of the selective spots. At Brooklyn Latin, for instance, girls were given 45 percent of the seats. Likewise, Stuyvesant and Brooklyn Tech had only 40 percent girls. At the High School of American Studies, girls make up 47 percent of the student body. Nor is it wholly true that some of these schools are junior versions of MIT. Despite its name, Brooklyn Tech offers sequences in social studies and law and society. The Bronx High School of Science boasts of its theatrical productions and its speech and debate teams. So it remains to wonder about the male preponderance, even at schools that stress Classical Latin and American Studies.

The answer lies in the entrance test, which governs admissions to all the schools I've mentioned. Its two sections, verbal and mathematics, are given equal weight. Since boys and girls score equally well on the verbal part, having the highest totals hinges on the mathematics segment. Eighth-graders who are drawn to American Studies or Classical Latin still must confront fifty questions like the one below, and answer each correctly in ninety seconds.

For a certain board game, two dice are thrown to determine the number of spaces to move. One player throws the two dice and the same number comes up on each of the dice. What is the probability that the sum of the two numbers is 9?

(a) 0 (b) 1/6 (c) 2/9 (d) 1/2 (e) 1/3

Correct answer: (a)

At this point, I'll only reiterate a question I've been posing: why must incipient scholars be made to show what they know in ninety seconds? So long as this rigidity remains, many of our most thoughtful girls and women will be denied places at our best schools. I'll have a suggestion about this at the end of chapter.

MISMEASURING MERIT

Ensuring opportunity has always been the American ideal, and there have been enough success stories to keep this hope alive. Yet recent years have raised concerns that all too many talents aren't being discovered or developed. Millions of young Americans seem destined never to realize their full potential.

If there's an organization that wants to be perceived as engaged in this sphere, it's the National Merit Scholarship Corporation. Based in Evanston, Illinois, the NMS sponsors a purported national talent search, funded by corporations eager to display a social commitment. Among its several hundred donors are well-known firms including Boeing, McDonald's, Southwest Airlines, and Lorillard Tobacco. Together, they provide the NMS with an annual budget approaching $50 million, cushioned by assets of $148 million.

The NMS says it aims to identify the nation's ablest high school pupils and assist them with their college bills. Schools across the country are urged to have their students enter its competition, in hopes of garnering money and recognition. Each year, some 1.5 million high school juniors from every state respond to this call, making it the most extensive enterprise of its kind.

The corporation emphasizes that all its scores, rankings, and awards are bestowed "without regard to gender, race, ethnic origin, or religious preference." In other words, its competition purports to eliminate bias for or against any social groups. At least, this is what the NMS says, and on this score it's in line with most other institutions that show a public face. (As it happens, geography is considered: candidates from Minnesota face a higher bar than those in Mississippi.)

Courts have long tussled with charges of discrimination. Since evidence of intent is hard to come by, an alternative is to see if a *disparate impact* ensues from certain policies or decisions. A college or an employer may truly believe it is looking only at merit, yet maybe something in its processes leads to finding that applicants of, say, Swedish origin lack the desired abilities. So even when there is agreement that merit is desired, the problem may be how that attribute is being measured and conceived.

In 2013, the most recent year for figures at this writing, the NMS began with 1,574,439 entrants, from which it culled 16,276 for public recognition. In the next few pages, I will be comparing the gender composition of the general pool with those whose names are publicized. The vehicle the NMS uses is the PSAT, which is basically a practice version of the SAT, intended for pupils in their high school junior year. As it happens, the PSAT provided a gender breakdown for those taking its 2013 test. Nationally, it was 740,969 boys and 833,470 girls, basically the same 47:53 gender ratio as for college graduates. It also makes gender figures available for each state. Yet the NMS itself never mentions that girls predominate in this pool. Perhaps there are reasons for this reticence.

The culled group—the 16,276—needs some explaining. They are called "semifinalists," since they will be subjected to more winnowing before the ultimate recipients are announced. (Indeed, only about 45 percent of them made that cut.) For reasons it does not explain, the NMS refuses to release a full list of its actual winners, let alone a gender distribution. All it provides are fifty separate state lists of aforementioned semifinalists, giving names and high schools. As a result, its semifinalists are the only group that anyone outside the NMS can examine. Still, even if many of them didn't make the winners' circle, they were all high scorers. I might add that this provides a good case study in statistics: do the best you can with what you have available.

The NMS likes to refer to its winners as "scholars," so it's appropriate to ask how they are chosen. The sole criterion is to score in the top one percent on the year's PSAT. Like the SAT, this test uses only standardized questions with predetermined answers, set in a multiple-choice format, overseen by a fast-ticking clock. For its scholars, the NMS simply asked the PSAT to print out the names of the 16,276 highest scorers, something easily done without human intervention. No interviews or essays are involved; there is no scrutiny of grades or supportive letters; nor does family financial standing figure explicitly in their selection.

So what are the gender numbers with this top PSAT-scoring 1 percent? Since it's ostensibly our academic elite, it would be interesting to know its composition. The NMS could easily have asked the PSAT for this data, since the students note their gender on a cover sheet before they start the test. Whether or not the NMS asked, they don't tell. NMS issues many multipage reports, largely laden with statistics, but at no point does it record how many in its 16,276 winners were boys and how many were girls.

So I asked Eileen Artemakis, the information director for the NMS, if she would give me some recent gender data. Here were her replies, along with my follow-up questions:

EA: *We do not compile data on gender or racial/ethnic origin because it is not relevant to the program since it is not used in the selection process.*

AH: *Does or doesn't NMS make—and examine in-house—a gender breakdown of awardees?*

EA: *NMS has in the distant past reported gender distribution. We no longer report this information because it has been misused and misinterpreted by others in the past. We expect that this will settle things for you.*

AH: *So you are admitting (or not denying) that you make gender tabulations in-house, but that you choose not to release them?*

EA: *[no response]*

Now some more information about the PSAT. The test has three parts—reading, writing, and mathematics—which are given equal weights and together make up each student's overall score, which can range from 60 to 240. Since the NMS is interested only in the very top scorers, I examined the gender distributions in the top tiers of PSAT test takers on each of the three parts. I found that the gender results were just about even in reading and writing. In reading, 0.9 percent of the boys and 0.8 percent of the girls were at the top. In writing, 0.8 percent of the girls and 0.7 percent of the boys were. So if only those two sections were used, the genders would end up with virtually identical representations in the NMS awards.

But—and I'm sure you know what's coming—this was not at all the case with the mathematics section. There, 2.5 percent of the boys scored in the topmost tier, but only 1.1 percent of girls did. Put another way, of the 27,954 PSAT pupils at the semifinalist level, almost exactly two-thirds—66.9 percent—were boys. If the percentages I cited in the foregoing paragraph are transposed into points, the boys averaged 41 against the girls' 27, which works out to a 52 percent edge. Given this hefty mathematics boost in the boys' total tallies, it's plausible to conclude they will outnumber the girls in the NMS's top-scoring group. Here, for the record, is one of the PSAT mathematics questions, to be answered correctly in seventy-nine seconds.

A 19-liter mixture consists by volume of 1 part juice to 18 parts water. If x liters of juice and y liters of water are added to this mixture to make a 54-liter mixture consisting by volume of 1 part juice to 2 parts water, what is the value of x?

(a) 17 (b) 18 (c) 27 (d) 35 (e) 36

Correct answer: (a)

Since the NMS refuses to divulge gender designations at any level, I pressed on myself. Because I was working on my own, I settled for examining a single state. I chose Ohio, which seemed sufficiently diverse as a sample. I set the stage with overall figures for the PSAT, which reported that 48,440 Ohio students took its 2013 test, and that 53 percent of them were girls. That was the easy part.

I then turned to the NMS list of Ohio semifinalists. There were 616 of them, in this case the top 1.3 percent of those in Ohio taking the test. Since the NMS wouldn't publish total gender figures, I made my own try at a count, based on first names.

About three-fourths of their names could readily be identified by gender. The remaining quarter either had national origins unfamiliar to me or were androgynous (Taylor, Morgan). After assigning them the same gender proportions as those I could identify, my Ohio semifinalist distribution came to 327 boys and 289 girls. So whereas girls made up 53 percent of those taking the PSAT in the state, boys accounted for 53 percent of the tier that determines NMS awards. Thus those extra mathematics points paid off handsomely for the boys. I have no reason to believe that this advantage is not the case nationally.

At one point, the NMS was helpful, even if indirectly and inadvertently. Its annual report listed the college majors their winners said they were contemplating. Over half of them (55 percent) clustered in just six fields: biology, business, computer science, engineering, mathematics, and physical sciences. But further figures I obtained from the College Board showed that girls were only half as likely to choose these fields. I consider this at least circumstantial evidence of the bias in the NMS's gender outcome.

What constitutes merit more generally—and how best to assess it—is a crucial question. As should be clear by now, the key component in the NMS metric is mathematics. As has been seen, problems like the one depicted earlier decide which of 1,660,047 entrants will be deemed most meritorious. A possible explanation is that the NMS has decided to favor students who seem to be adept in STEM-related skills. If that's its disposition, it should be openly stated, along with acknowledging that this bias lessens the chances of its awards going to girls.

The National Merit Scholarship Corporation is funded chiefly by business firms, presumably as a public and social service. At least in theory, these companies profess a desire to give women a full chance to discover and develop their talents. A good start would be to advise the NMS to reexamine its protocols so that women have the same opportunity as men in a competition purporting to help young people pursue higher education.

From the outset, this book has agreed that mathematics is an honorable calling and deserves an esteemed place in the pantheon of learning. But it is something else again to make it central to entering the citadel of merit. (At no point does the NMS show it deems sculpture or sociology, or a gift for political science or diplomacy, to be serious or necessary talents.) If the NMS model is embraced—and it increasingly is—we will accept a constricted conception of who among us has abilities and strengths. It is not just a nation's mind that's in the balance. At stake is the kind of people we want to be.

WHEN WOMEN OUTSCORE MEN

Despite my misgivings about the standardized industry, I have to concede that the SAT offers a huge repository of information, which it culls from the millions of students who take its tests. Along with detailed data on scores, it releases information the students provide about themselves. So I asked the SAT to tell me about two groups of high school seniors who took its test. For the present, I'll simply designate them as CM and AW. The table on this page gives the average mathematics scores of the students

SAT: Scores and Status

CM		AW
544	Average mathematics score	566
9%	Scoring 700 to 800	16%
63%	Parents college graduates	56%
$77,000	Average family income	$42,000

in the two sets, along with how many scored in the top tier, and some facts about their families.

What's readily apparent is that even though more in the CM group have parents who are college graduates and have visibly higher incomes, they still score lower in the mathematics part of the SAT. The AW group does better, despite having fewer home-based advantages.

Thus far, I've been coy about identifying the groups. So now I'll confirm what you must have guessed: "M" and "W" stand for men and women, while "C" and "A" refer to Caucasian and Asian American designations. Of course, it's always possible to find some women who outscore men. On the 2014 SAT, 46,400 women scored over 700 in mathematics, which meant they outperformed 706,189 men who fell below that mark. Marathons also have women who finish ahead of the general run of men. In this case, AW denotes the entire pool of Asian women who took the SAT, not just those performing especially well, while CM includes all the Caucasian men who took the test. The average performance of all Asian American women is twenty-two points higher than the average for all Caucasian men. But since the average for all women falls below the average for all men, we must ask if there is something about the experience of being Asian American that leads these women to turn in a superior performance on a national test.

In fact, I have no problem saying that. After all, it isn't that Asians have a calculus chromosome. The cause, rather, is that their culture stresses academic achievement, respect for parental authority, and upholding the honor of one's family more than a typical Caucasian American family does. As a result, Asian American women not only surmount the SAT's gender-based barriers, but go on to outscore a large phalanx of Caucasian men. And since we've ruled out genetics, it's worth pondering what Caucasians generally might have to do in order to improve their scores. It's easy enough to ask for better study habits and academic attitudes. But that would involve altering the predilections and priorities of an entire ethnic cohort, a transformation not easily achieved.

THE WEEKEND

"Why Can't a Woman Be More Like a Man?" demanded Professor Henry Higgins in *My Fair Lady.*

One solution to the multiple-choice gap would be to invest in intensive tutoring for girls and women, to teach them how to act more like men and boys when taking the standardized tests. After all, it could be argued, tests like these are here to stay, so it's best to adapt to their regimens and succeed on the terms they set. Going forward, girls would be advised to be less cautious and contemplative, and not mull over ambiguities, or their suspicion that several answers are possible. Along this route, they would recast how they use their minds, the better to size up problems in 75 or 79 seconds. Since just finishing the tests requires a proclivity for taking risks, women would be urged to learn some lessons from men.

I regard this as a wrong turning, since it gives velocity priority over the content of what's supposedly being tested. I want to close this chapter on a constructive note, in hopes of resolving its central quandary. If we assume, as we should, that there's a lot of mathematical potential out there, how best to uncover those abilities? Since for much of the population, the chief obstacle to showing what they know has been a stress on speed, let's see what happens if we eliminate that hurdle.

Let's select a sample of high school seniors of both genders and give them an alternative to the current regimen. They will be allowed to take mathematics questions from the ACT or SAT home with them and to work on them over an entire weekend. They will be encouraged to take their time, not try to finish them at a single sitting, to return periodically to review their answers and decide if they want to change them—much the way real people engage with assignments in the real world. Needless to say, this will be on an honor system. Students will be asked to complete the assignment wholly on their own, not to call friends, or consult references or other sources. At the risk of undue naïveté, I'd like to think that most would honor this request.

Then let's analyze the results. It seems safe to assume that both genders will do better with more time. But whether this approach will raise the women's scores more than it will the men's is one of the hypotheses we'll be checking. Another would be to see how many more of each gender ascend to the 700–800 tier, which is now achieved by only 9.6 percent of the boys and 5.2 percent of the girls.

I'm not prepared to predict that our weekend girls would end up with scores paralleling those of the weekend boys. After all, the ticking clock is not the only facet of these tests that isn't suited to women. The aim of the experiment is one we should all applaud: to give people a chance to show what they know and can do. If having more time accomplishes that, I can't see why anyone should object to such a finding. I find it amusing that the ACT and SAT have never done a study akin to this. Might it be they'd rather not have the results on record?

6

Does Mathematics Enhance Our Minds?

Does studying algebra enhance our intelligence? A lot of people think so, and their arguments often seem compelling. It's surely true that mathematics challenges the mind. Simply gaze at students struggling with geometric proofs. We literally see their fists clench and brows furrow, as they press their capacities to the fullest. We can all agree that mathematics makes mental demands.

At a loftier level, mathematicians are viewed with awe, a grand Olympus of sages and scholars. They seem blessed with a higher order of intellect, enabling them to enter realms beyond the reach of other mortals. So it's not surprising to hear it said that studying mathematics will broaden our minds generally, girding us for challenges beyond the discipline itself.

Auguste Comte, a founder of modern philosophy, believed this was so of mathematics' basic branch. "Algebra strengthens the mind," he said, "and enables it better to master studies of a different nature." In our own time, the National Council of Teachers of Mathematics affirms that "a person who has studied mathematics should be able to live more intelligently than one who has not." Likewise, the National Research Council, an esteemed public agency, claims that mathematics instills *procedural fluency, productive dispositions, conceptual understanding, strategic competence*, and *adaptive reasoning*. In an era that rewards analytic skills, mathematics is held as a crucial key to logical thinking and revealing the promise of mortal minds.

This chapter will challenge these suppositions. A more dispassionate look will show that the arguments I have cited are largely wishful thinking and far from proven propositions. Rather, as will be seen, the claim that studying mathematics uniquely instills desirable modes of thought is built on premises that have never been verified and erode on closer examination.

I readily grant that mastering mathematics calls for special powers of reasoning. I like the way Alice Crary and W. Stephen Wilson, who work jointly in mathematics and philosophy, put it: "Mathematics demands a *distinctive* kind of thought." (I've added the italics.) Obviously, you must use your mind to progress toward a geometric proof, whether it's seventh-graders parsing angles or professors conducting research on ellipsoidal coordinates. But all sorts of endeavors can call for hard thought, from demystifying postmodern poetry to perfecting haute cuisine. It has yet to be shown that adults who were once made to master Pascal and Pythagoras end up more generally thoughtful than classmates who sharpened strengths in other areas.

In fact, scholars themselves raise serious doubts about whether modes and methods instilled in one area can be applied elsewhere. I'll quote three:

> There is some question about whether the training to think in one sphere carries over to thinking in another. One may be inclined to believe that it does, but one could not prove that this is so.

> Being good at one mental skill does not necessarily train the mind to be skillful in other domains. This is one of the most solid findings in psychology, confirmed and reconfirmed many times.

> Improvement in any single mental function rarely brings about equal improvement in any other function, no matter how similar.

The first is from Morris Kline, a renowned New York University mathematician. Next comes the University of Virginia's E. D. Hirsch, widely known for his lists of what every cultured person ought to know. The third is Edward Lee Thorndike, an eminent Columbia University psychologist, who issued his statement back in 1923, following his finding that having studied Latin didn't improve people's literary skills. No one I know will contend that immersion in anthropology will make you an expert astronomer. Or that setting swimming records will take you to the top in lacrosse. Mathematics is the only endeavor I know of that claims to bring universal mastery.

There can be no denying that reasoning and logic are foundational to mathematics. At the same time, I'm waiting to be shown that agility with polynomials produces sharper insights on other topics. Can *quod erat demonstrandum*, even if exquisitely structured, help us resolve whether deciding to end a pregnancy is taking a life, or if the national interest will be served by invading another country?

The proofs associated with mathematics are schematically structured, with each step numbered or similarly designated. Bertrand Russell has called this logic "cold and austere," not only in avoiding emotions, but as a pursuit of highly disciplined minds. Deliberation in the wider world is seldom so ascetic. While debates about abortion and invasions may deploy facts and figures, it hardly needs saying that emotions, cultural values, and other unquantifiable factors come into play. In sum, it's not easy to enumerate human problems that can be solved with an amenable *QED*.

Morris Kline isn't the only mathematician to doubt the transferability of mathematics' reasoning skills. Peter Johnson, who teaches at Eastern Connecticut State University, alludes explicitly to algebra, but his remark could apply to the entire discipline: "There appears to be no research whatever that would indicate that the kind of reasoning skills a student is expected to gain from learning algebra would transfer to other domains of thinking or problem solving or critical thinking in general." He is seconded by Underwood Dudley of DePauw University, who writes: "To assert that mathematical training strengthens the mind is as impossible to prove as the proposition that music and art broaden and enrich the soul."

As a matter of fact, in most of our lives, we move along without proving propositions, but try to get by with a fair degree of certainty (like believing that music and art are important for us). In this and later chapters, I will propose other—and more effective—ways to advance the kinds of reasoning so often attributed to mathematics.

CRITICAL READING AND COLLEGE HISTORY

I'll start with two modest studies of my own, which I undertook because reliable research is sparse. The first draws on the Scholastic Assessment Test, which is a key factor in college admissions and comes close to being a national IQ test. I wanted to know about the SAT reading scores for those students who excelled in mathematics. In other words, if you are very good at mathematics, are you also likely to be a very good reader?

I asked SAT if they would prepare a spreadsheet showing how students who had the highest mathematics scores fared in the reading section. I wanted to find out if the intellectual habits absorbed by successful students of mathematics carried over into scrutinizing paragraphs of prose. As in any statistical study, some choices had to be made. In this case, I decided to limit my analysis to students who gave their race as being either black or white. My concern was that many Asian American and Hispanic students are new to this country, which might affect how they performed on the reading section.*

In fact, 5.4 percent of the combined black-plus-white pool scored 700 or higher on the mathematics portion. These high scorers are arguably among our finest young mathematical minds, at least as gauged by a test like the SAT. So I must confess my surprise when I found that only somewhat over a third—36 percent—of this mathematical elite also attained 700 or higher on the critical reading part. This hardly supports the view that reasoning honed in mathematics segues into other areas.

I then decided to run the test in reverse, looking at all students who scored 700 or better in the reading part. And here came another surprise. Fully 44 percent of them also recorded 700 or better in mathematics. In a word, these humanities mavens surpassed their mathematics classmates when it came to excelling in another testing terrain. So these SAT figures suggest that literary proficiency is more likely to be accompanied by mathematics achievement than the other way around. Humanities scholars might well argue that if we want to improve agility in algebra, more time should be spent discussing novels and poems.

Next, I conducted a modest study at Queens College in New York City, where I teach. All of our freshmen must take a placement test, which has a mathematics part. Here also I wanted to find out how students' scores on this section compared with their performance in a general history course, which a fair number of them had elected. I did this because history straddles the humanities and the social sciences, requiring a range of skills, including reasoned reflection and evaluation of evidence.

The mathematics test had a 60 to 100 scale, while history's grades were the customary A+ down to D and F. In the graph

---

*Altogether, 88 percent of the black students and 93 percent of the white said that English was the first language they had learned. For those identifying as Hispanic and Asian American, the comparable figures were 31 and 28 percent.

below, each dot represents where students' mathematics scores and history grades intersected. As can be seen, they scatter all over the graph. In the end, those who scored well in mathematics did no better in history than their classmates with low mathematics scores. Indeed, a textbook correlation between mathematics scores and history grades was just about zero (+0.0699), which is about as close as one can get to there being no relationship at all.

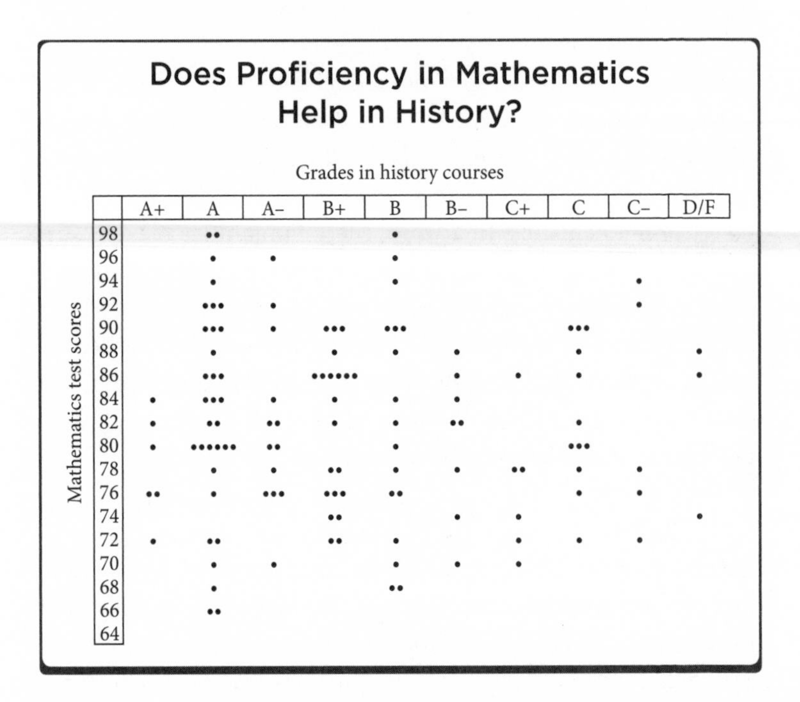

REASONING AND REPRESSION

It is widely assumed that human inquiry will flourish only if the life of the mind is honored and ensured. Indeed, such an atmosphere distinguishes societies we call free from others not so blessed. So it seems apposite to ask whether this stricture holds for mathematics, as it does for other disciplines. This issue is especially germane in light of claims that mathematics' modes of reasoning can enrich our understanding of the entire human and social condition.

This occurred to me when I saw the results of the 2013 International Mathematics Olympiad, where 527 students from ninety-seven nations competed in Santa Marta, Colombia. (Each country is asked to enter a six-member team. As the numbers show, not all could or did.) These were the highest-scoring teams:

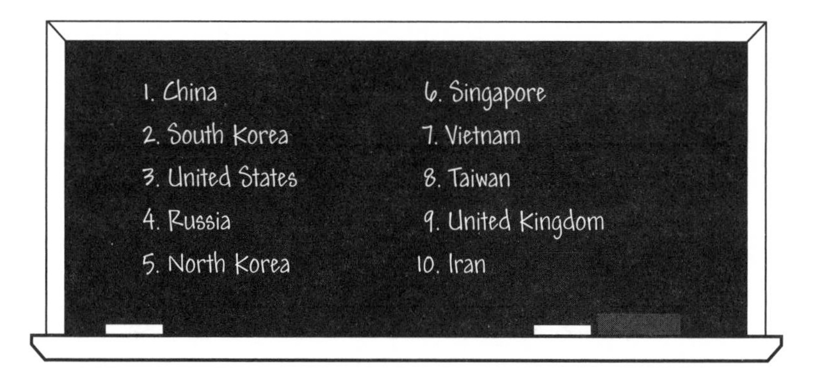

Several things are striking about this list. One, certainly, is that Asian teams took six of the top places. In addition, half of the entrants from two more countries—the United States and the United Kingdom—had names that suggested Asian origins. There's no doubting that these front-runners are incredibly talented young people. While some of their countries may stress rote learning and standardized testing, the few who reach Olympiad ranking can't simply be dismissed as agile automatons. A glance at the problems the Olympiad poses shows that they call not just for correct answers, but ingenious routes to those results. Top scorers must be intellectually creative, at least as mathematics construes that talent.

What also caught my eye is that half of the winning countries are not known for safeguarding, or even allowing, unfettered freedom of inquiry. China, Russia, and Vietnam make life difficult for dissenters, barring them from recognition and rewards.

And it hardly needs saying that Iran and North Korea rank high as repressive regimes. Even in Singapore, caution is a prudent path for those seeking approved careers. In sum, at least half of the Olympiad victors were raised amid restraints on what may be said and taught and learned.

Granted, dictatorships can't ban what goes on within people's heads. Still, when public discourse is constrained, the overall environment is affected. It takes special, and often hazardous, efforts to think beyond the state's parameters. This noted, the list of Olympiad winners is evidence that young people reared under constricting conditions are able to develop world-class mathematical minds.

So what can we conclude about mathematics, when its brand of brilliance can thrive amid onerous repression? One response may be that the subject, by its very nature, is so aloof from political and social reality that its discoveries give rulers no causes for concern. If mathematics had the power to move minds toward controversial terrain, it would be viewed as a threat by wary states.

Scholars in other fields, notably the humanities and social sciences, have always attested that their studies can flourish only when open inquiry is maintained. Biologists will tell you that they still worry about obstacles to teaching evolution, while scholars who focus on the climate say they often find themselves on the defensive. All can cite cases where outsiders with ideological agendas have impugned research in their fields. But mathematicians appear to be an exception. We never hear them complaining that what they teach and study has roused political reactions. Apart from debates over curricular protocols—which I will consider in another chapter—they can't point to anything they do that might stir outside ire. At most, some will say they would like more funding for their research, blaming its paucity on anti-intellectual tendencies. So I've found myself asking: is there such a thing as a *freedom* that mathematicians need? Is there a milieu that scholars and students of mathematics must have if they are to perform at their intellectual best?

The Olympiad results suggest that such conditions are just as available in Iran, China, and North Korea as in the United Kingdom and the United States. It may be argued that were such societies more open, they would produce even better mathematicians. But that's hard to assert when China already leads the league, as it has more often than any other country. Of course, mathematicians in freer societies can also express themselves in other realms, as many do. But this doesn't mean that their contributions, say, to conic sections add greater weight to their opinions on gay marriage or gun ownership.

THE PURSUIT OF PROOF

The great goal of mathematics is to *prove* propositions. To start with what had been a *conjecture* or a *hypothesis* and devise steps that turn it into a *truth*. Establishing such proofs requires a great deal more than making a reasonable case. For a mathematical theorem to be proved, its steps must be so persuasive as to secure what is essentially unanimous agreement from members of the profession. Today, proofs can run to hundreds of pages, frequently requiring the lightning speed of computers.

I readily agree that the quest for mathematical proof calls for unusual imagination, a willingness to question prevailing premises, and to explore unexpected alternatives. So, pursuing proofs in mathematics is an exalted exercise. I like how Roger Penrose put it:

> A proof, in mathematics, is an impeccable argument, using only the methods of pure logical reasoning, which enables one to infer the validity of a given mathematical assertion . . . from axioms whose validity is taken to be self-evident.

A familiar example would be Euclid's III.35. Euclid's feat was to prove that we can construct a square whose area is equal to the area of a given rectangle. I won't delve into the details, but will simply note that he drew a circle around a rectangle, and then used the circle to measure the sides of an equal-area square. From that came the equation $ab \times bc = fb^2$, after which he felt he could append *quod erat demonstrandum* at the bottom of the page. How did he do it? Penrose just told us: "pure logical reasoning." Once Euclid announced his proof, other mathematicians gave it careful scrutiny, as should occur in every science. They felt free to hunt for discrepancies, inconsistencies, or other flaws. Yet in Euclid's case, the consensus persists that his III.35 has been proven beyond a doubt.

What should also be underscored is that the arguments and axioms alluded to by Penrose are deployed in and for a realm almost wholly detached from earthly experiences. This being so, I have come to conclude that learning how mathematicians pursue proofs has little or no connection to the pursuit of proof in lives the rest of us lead. For this reason, I will propose that there are other, more fruitful, ways to understand the meanings and processes of real-world proof than from the study of mathematics. These alternative approaches can challenge our mental powers no less profoundly than geometry and algebra, and relate far better to our personal and public lives.

SCIENTIFIC PROOF: UPDATING DARWIN

As has been noted, in mathematics and the sciences, advancing a theory or theorem rests not just on the acuity of one's analysis, but on obtaining the agreement of one's peers. In the sciences, even when that verdict is obtained, it is more tentative than in mathematics. That's because natural and physical scientists confront what William James once called a "blooming, bustling confusion," whether it's out there in the universe or deep within ourselves. Everything from astronomy to zoology is so vast and interconnected that we'll never find all the pieces let alone decipher the overall puzzle itself. Mathematicians may assert, as

Simon Singh has, "once something is proved, it is proved forever." Scholars in other sciences cannot share that certainty.

So what do we mean when we say that a scientific theory has been proved? Let's start with first asking what it means to call a body of knowledge a "theory." A good case is Charles Darwin's insights, which he chose to describe as "natural selection" or simply "descent." Most modern scientists now call it "evolution," as do those hostile to Darwin's ideas. As is well known, some critics want Darwin's theory removed from classrooms, on the grounds, as they put it, that "it's just a theory." This phrase, plus the sneer behind it, is meant to convey that the concept of evolution is a maelstrom of shaky speculations.

Adherents, on the other hand, grant that Darwin isn't the final truth. And for an honest reason: in the sciences taken together, finality is rarely reached. Thus, natural selection is the most coherent analysis we currently have of how life takes the forms it does. It's the theory that best explains what we observe. So to say that it's been "proved" states only that Darwin provides the most satisfying answers we have thus far for many of the existential questions we confront.

If we want students to grasp the process of scientific proof—and its sister, scientific truth—they could be invited to examine how Darwin's theory of evolution has itself evolved since he first started collecting notes on the H.M.S. *Beagle*. One way to start is to consider how natural selection, adaptation, and survival have been synthesized with our growing knowledge of genetics. To start, there was Hugo de Vries's focus on mutations, arising from his studies of hybrid flowers. Thomas Hunt Morgan, working with fruit flies' chromosomes, gave natural selection further amplitude. J.B.S. Haldane and Ronald Fisher created statistical structures for evolution, and Theodore Dobzhansky clarified how natural populations develop. But the genius of Darwin is that his conception can absorb these revisions and additions. Thus recent research on genetic mutations doesn't undercut Darwin. Rather, it enhances our understanding of our continuing "descent."

By using Darwin as a case study, we can consider how far his ideas have been validated, or even whether "proof" is the issue. Grappling with such questions will impart more about the quest for certainty than will relying on Euclid or Pythagoras. In fact, there's hardly any mathematics in Darwin's theory, which is itself worth pondering. There's no denying that many of the proofs in mathematics are breathtaking achievements. Even so, they don't help much in establishing truth in other realms.

LEGAL PROOF: BEYOND A DOUBT?

In my time, I've served on five juries. All were criminal trials, with two of them involving murders. I recommend such service to all adults, not only on civic grounds, but because it offers a rare educational experience. You learn a great deal about your fellow human beings, even gain insights into yourself, as you become part of a pursuit we like to regard of as ensuring justice.

More than that, I have come to conclude that being a juror can teach us more about proof than several semesters of geometry. A panel of citizens is asked to discover and decide if the evidence presented by prosecutors has *proved* the guilt of a defendant *beyond a reasonable doubt.* (In a civil case, juries can be smaller in size and need only find whether a *preponderance of the evidence* has been adduced for one party's side.)

Just as science is based on attaining a consensus about evidence, so must a group of citizens in a jury reach a common verdict. A lay panel of men and women is asked to muster a collaborative intelligence, to assess whether what they've seen and heard has met stringent standards explained by the judge. As we know, defendants are not expected to prove that they are innocent. Rather, officials of the state must prove guilt.

Here's a vignette from one of my trials. A police officer was sworn as a witness. He testified that he was heading home from duty, when he heard a shot from inside a house. He dashed inside, he told us, where he saw a body by the door, and a man standing, holding a gun. That man was now the trial's defendant, charged with murder. Even if the gun wasn't smoking—we learned that doesn't happen anymore—it's not often that a case seems so clear. But then we heard the accused man's defense.

He had been upstairs, he testified, when he heard gunfire on a lower floor. He dashed down, and saw the body and a gun alongside it. He admitted that he retrieved the weapon, since in his neighborhood it could fetch a substantial sum. Somebody else was the killer, his lawyer argued in the summation. Did we want to send an innocent person to prison? It was left to us, as jurors, to decide whether the prosecution had presented enough *proof* that the man before us had committed a murder. After a day of careful and sometimes contentious deliberation, we reached a unanimous decision.*

Students could be asked to simulate being on a jury. They could read the transcript of a trial, watch a courtroom video, even ponder a fictional case. Whichever method is used, they could be exposed to oral and physical evidence, the arguments of the attorneys, testimony of witnesses, along with instructions from a judge.

The challenge is to weigh evidence and decide if one side has *proved* its case by criminal ("beyond doubt") or civil ("preponderance") standards. Moreover, a trial is not a totally open forum. Jurors are told to assess only testimony and exhibits that have been deemed admissible during a trial. (Science has its own rules of admissibility.) If witnesses tell conflicting stories, jurors must weigh whether this may be due to fallible memories, partial perceptions, or even outright lying. Nor are jurors passive receptacles. They are expected to deliberate among themselves, sharing what might be varied interpretations of what they saw and heard at the trial. The presumption is that they are reasoning beings, able to weigh evidence, and listen and learn from one another.

There is no guarantee they will reach a unanimous verdict,

---

*As it turned out, we found the defendant guilty, with no reasonable doubts about our finding.

which is another lesson in the proof-seeking process. They may even find that the quests for truth and proof don't always intersect. After one case, I recall telling a friend, "We felt that the defendant was probably guilty, but we concluded prosecutors hadn't presented enough evidence to prove their case."

Am I saying that the quest for justice has much in common with mathematics? In many ways, it does. After all, both want to ascertain the truth, whether it's about vectors and angles or whether the man holding the gun has also pulled its trigger. In more than a metaphorical sense, jurors are asked to create multivariate models, which winnow and weigh the factors that will go into an ultimate verdict. Whether the terrain is evolution, a courtroom, or a classroom, what we're looking for is logical thinking and an ability to evaluate real-world evidence in searching for truth as closely as we can come.

But I'm also saying that mathematics has no special claim to logical thinking. Lots of endeavors—from serving on a jury to understanding competing views of evolution—require forms of logical proof. And some of these are arguably more valuable to students' lives than abstract mathematics.

This chapter is also intended to challenge mathematicians: to show how Euclidean proof shares ground with its legal and scientific counterparts. It's worth a trial.

VOICES

I would be curious to know if there have been any gold standard research studies that demonstrate that math courses actually lead to students having greater analytical prowess.

Where does the belief that everyone should take mathematics come from? Aren't there equally challenging ways to mold the mind other than doing mathematics?

There appears to be no research whatsoever indicating that the kind of reasoning skills students are expected to gain from algebra will transfer to other domains of thinking or to problem solving in general.

All the mathematics one needs can be learned in early years without much fuss.

VOICES

The argument that algebra is the only way to produce a critical thinker troubles me. As a social studies and literature teacher, I would like to believe that these courses do their part as well.

Reasoning mathematically may be a nice skill, but it is not relevant to most of life. We reason about many things: parenting, marriage, careers. Do we learn how to reason about these things by learning algebra?

I know people who can work out most of Fermat's Last Theorem but keep marrying the wrong women.

Are mathematicians the best thinkers you know?

7

The Mandarins

Given their impact, they are a relatively small group. Most of them are senior professors of mathematics at our top-tier universities, best known for graduate programs and advanced research. Others may be at less elite institutions but make their mark by holding office in scholarly societies, serving on public commissions, and pronouncing on the state of the discipline. So in addition to being erudite, they seek to exert influence.

I will be calling them "the mandarins," since they have much in common with ancient China's caste, not least for their aura of complacency and privilege. Lynn Arthur Steen, at St. Olaf's College, depicts them as "the mathematics power elite," playing on C. Wright Mills's trope for America's corporate overseers. Paul Halmos, who made his mark with ergodic theory, sees them as a "self-perpetuating priesthood." Either way, they are an academic coterie, whose members preside over how an entire realm of knowledge is defined, taught, and studied at every level.

But doesn't it make sense to defer to masters of a discipline? We leave vital decisions to experts like actuaries and neurosurgeons, or at least think twice before intruding our views. Every society has professionals who know more about some things than the rest of us. For my own part, I have no problem with paying brilliant men and women to think big thoughts within sequestered institutions. Indeed, I'm happy to see my tax money supporting research on Hilbert spaces and Wolstenholme primes.

So here's my concern. It's that the mandarins aren't content to stick to their scholarship. Rather, they take it as given that their intellects give them license to dominate much of our educational system and set priorities for the greater society. In a word, they have a cause, founded on the supposition that their realm reveals the human mind at its finest. Among some mandarins, mathematics is a secular ideology; for others, it is a cerebral theology.

For most adults, even those with reputable educations, mathematics—in contrast to arithmetic—is terra incognita. When the journalist Betty Rollins married a prominent New York University mathematician, like most spouses she sought to understand what her partner did all day. After a few hapless evenings, she consulted with other faculty wives, all of them with bachelor's degrees or better, and most with serious careers. "Don't try," they told her. "There's no way you'll get even halfway to first base." (The field is due for a companion movie to *I Married a Spy.*)

Of course, senior professors should decide what to teach in their seminars and the courses required in their fields. We call this academic freedom. But it's another thing for an unaccountable elite to lay down what every young person in the country will have to learn. As matters stand, mandarins are reflexively appointed to committees and commissions, which do not hesitate to dictate what should transpire in tens of thousands of classrooms, even prior to kindergarten.

TWENTY-SEVEN TOPICS

This mentality was evident in a 2008 report from a panel chosen by Margaret Spellings, President George W. Bush's secretary of education, which called for top-down changes in K–12 teaching. Filling 857 pages, and titled *Foundations for Success*, the report began by affirming that "sound education in mathematics across the population is a national interest." In fact, I support this sentiment, although for reasons other than economic primacy or military security. After all, mathematics can and should be studied for its truth and beauty, as are art and poetry. I'd like to think that aesthetics are also a "national interest."

In any event, I found myself concurring when the commission urged that all young people "develop conceptual understanding, computational fluency, and problem-solving skills." In a later chapter, I will offer suggestions on how such aims might be achieved. But after these preliminaries, the panel's agenda became evident. Its members concurred that all "students must be prepared with the mathematical prerequisites for major topics of school algebra," starting in the earliest grades. It then listed twenty-seven of these topics, ranging from *rational expressions* and *binomial coefficients* to *quadratic polynomials* and *logarithmic functions.*

Some insights can be gained from looking at who was on the Bush panel. Fifteen of its twenty members were or had been senior professors at universities at the level of Harvard, Berkeley, and Cornell, which emphasize research and doctoral degrees. Two foundation officials and an administrator of a suburban college also had seats. Almost as an afterthought were a retired elementary school principal and a solitary middle school teacher. In the final reckoning, the university scholars outnumbered the lone school teacher by a ratio of fifteen to one.

This explains their stress on advanced algebra for everyone. They see this as a step on a road to the realms that are the mandarins' personal preserve. In truth, only an infinitesimal fraction of our eleven-year-olds will end up in doctoral seminars. But to ensure that just this handful will be properly groomed, the panel enjoined the nation to prescribe heavy doses of algebra for every child who walks—even toddles—through a school door. W. Stephen Wilson of Johns Hopkins urges "laying the foundation for college readiness in mathematics early, by grade six or seven." On first reading, this may seem reasonable. But we should be ready for the fallout if "college readiness" requires that all four million of our seventh-graders master a stringent mathematics sequence.

Foundations for Success had a deeper impact than was noticed at the time. Its mantra of advanced mathematics formed the groundwork for a project that came to be called the Common Core State Standards, which would be embraced a dozen years later across most of the country. As I'll be suggesting in a later chapter, no one can object to efforts aiming to raise standards. But imposing the mathematics of "college readiness" on everyone will derail the academic careers of huge pools of our young people.

SIXTY-EIGHT SKILLS

The mandarins are active on many fronts. Another of their campaigns seeks to warn students about the kind of mathematics they will encounter once they get to college. The message came in a report called *Standards for Success*, sponsored by the American Association of Universities, which represents the nation's most research-oriented institutions, and is funded by the Pew Charitable Trusts. Some two dozen institutions, ranging from Harvard and Stanford to Indiana University and the University of Michigan, joined in specifying the skills students should have if they wanted one of their degrees. (For reasons not given, even highly regarded liberal arts colleges like Swarthmore and Carleton were not invited to take part.)

Copies of *Standards for Success* (abridged to *S4S*, perhaps to convince youthful readers that it was hip and cool) were sent to some 20,000 high schools, along with CDs highlighting the report's findings. Specifically, the report sought to alert students to "the quality of work that AAU-university professors will be expecting of freshmen in entry-level courses." It added—ominously—that high school graduates "are often surprised by the knowledge and skills university professors expect of them."

To prevent such shock on the mathematics side, *S4S* lists no fewer than sixty-eight skills it says should be absorbed in high school, the better to ensure readiness for a rigorous freshman year. Here's a sampling of what the professors say they will be expecting of students taking a mathematics course, as most must do.

<> Understand a variety of functions (e.g., polynomial, rational, exponential, logarithmic, and trigonometric) and properties of each.

<> Understand the relationship between a trigonometric function in standard form and its corresponding graph (e.g., domain, range, amplitude, period, phase shift, and vertical shift).

<> Understand the basic shape of a quadratic function and the relationship between the roots of the quadratic and the zeroes of the function.

The message is that this and more is what you'll need to know if you want to pursue an undergraduate degree at Harvard or Indiana, as well as at other sponsoring schools including Rutgers and Rice. Even students hoping to major in art history or anthropology should come girded in vertical shifts and quadratic roots.

As noted, the report says it is telling students what "university professors expect of them." Yet, in fact, young people who embark on higher education begin in *colleges*, even if these are located within universities. That is, freshmen follow a separate curriculum, which leads to bachelor's degrees. For this reason, it is in order to wonder whether the professors who have been pronouncing on standards and success will actually be their teachers.

It would be revealing, for example, to learn how much time senior scholars at Harvard and Stanford and Michigan devote to undergraduate instruction. Nor do I mean lecturing two hours a week to a huge hall. How many personally take on a section of 101 or a freshman seminar? Some answers will be provided later in this chapter.

Sadly, I've come to conclude that the mandarins are less concerned with individual students, or even the nation's mental wellbeing, than filling seats in their own PhD programs. In many ways, we've been on this route before. A little over half a century ago—in 1957, to be exact—the then Soviet Union became the first nation to reach outer space, launching a satellite it called Sputnik. This feat alarmed the United States. One of the first reactions was to decide that mathematics teaching must be revamped, so we could regain hegemony over our primary adversary. Not surprisingly, at the forefront of this effort were university mathematicians. American pupils, they complained, were being taught computational techniques, but failed to appreciate theoretical principles. In fact, this lack was also evident among their own doctoral candidates. So they urged that mathematical theory be started in elementary grades.

What soon came to be called "the New Math" was introduced. Teachers were sent to workshops to prepare for a Supra-Sputnik era. Parents saw their sons and daughters given assignments in set theory, vector space, commutative and distributive laws, not to mention the ubiquitous Venn diagrams. Books were sold to moms and dads, with tips on fathoming their children's homework. The New Math may have looked fine in university conference rooms, where it originated. But it soon became clear that it was unworkable, even for teachers who had had strong mathematics groundings in college. Suzanne Wilson of Michigan State University, in a classic analysis of the debacle, concluded what should have been obvious at the time. The New Math, she said in her definitive book *California Dreaming*, "failed because it was led by mathematicians, not mathematics teachers."

HOW NOT TO HELP K-12

A sign that ours is a modern age is that we have more theoreticians than ever in the past. It often seems that fewer people are doing something tangible, while entire professions are devoted to ideas, analysis, and explication. Almost daily, classroom teachers receive new directives concerning technology, testing, and more stringent mathematics standards. Yet all too often, these edicts are oblivious to the range of capacities among the huge range of young students our teachers are asked to educate. Here's an apt illustration.

Not long ago, and with much fanfare, the Irvine branch of the University of California announced that it had been given $2.5 million to establish a "Teaching Chair in Mathematics," designed to prepare mathematics teachers rather than mathematicians. For many reasons, there was need for such a gift. While the state's two university systems awarded some 120,000 bachelor's degrees each year, a mere 200 of these graduates were receiving accreditation as high school mathematics teachers.

To try to induce more undergraduates to enter so vital a career seems a praiseworthy goal. All the more, since the stated aim of the Irvine donation was to enhance "excellence in teaching." Still, it carried a specific condition. The money had to be used to attract a "distinguished research mathematician," who would be asked to take a leadership role in the preparation of science and mathematics teachers for California's public schools.

I hope I will be forgiven for thinking something is amiss in believing that a "distinguished research mathematician" would be a person to contribute to the training of K–12 teachers. Whatever such a mandarin's renown, there is no evidence at all—and much contrary—that this eminence could help make mathematics more appealing to pupils aged five through seventeen. Scholarly research in mathematics is in an altogether different galaxy from even the most rigorous courses at highly selective high schools.

My own proposal would be that the chair be awarded to a tenth-grade teacher from, let's say, a Fresno or Santa Rosa high school. I am sure there would be many excellent candidates, who are known for stirring enthusiasm among students hitherto indifferent to polynomials. But that such an individual might be named to the Irvine chair is about as likely as an imminent proof of the Riemann hypothesis.

DISAPPEARING STUDENT BODIES

Here's a not-so-small irony. Given all the advice the mandarins dispense about the state of education, they have been conspicuously unsuccessful in attracting young people to their discipline. Much of the story is told in the table below, which compares the number of mathematics degrees awarded in 1970, a little over a generation ago, with those granted in 2013, the most recent year for figures at this writing. The raw numbers in the table show a sharp drop in mathematics majors, from 27,135 to 17,408. But

Disappearing Degrees

Mathematics Awards to U.S. Citizens

1970		2013
27,135	*Bachelor's*	17,408
3.4%	of all majors	1.0%
5,145	*Master's*	1,809
2.5%	of all majors	0.3%
1,052	*Doctorates*	730
3.5%	of all majors	0.5%

the real decline is much greater, because the number of bachelor's degrees overall has doubled since 1970. So it's best to look at where mathematics stands relative to all bachelor's awards. By that measure, its majors in 2013 accounted for precisely 1 percent, less than a third of their share in 1970. It's hard to find another academic discipline that has plummeted this far. And, as can be seen, the drops in graduate degrees have been equally dramatic.

In short, students are eschewing what was once a leading discipline. Of course, the mandarins are ready with explanations. Here, summarized, are some I've heard when I've raised the subject in conversations.

- *There has been a movement away from majors that stress learning for its own sake, including the pure sciences like mathematics.* This is largely true. Even so, the drop in mathematics has been steeper than in many other fields.
- *Computer science has supplanted many science majors, since it is reputed to offer entry to successful careers. The field of study barely existed in 1970, but awarded 50,962 bachelor's degrees in 2013.* Fair enough. So here's what I'd like to ask. If computer science weren't available, how many of those now choosing it would have chosen mathematics? Designing hardware and coding software are essentially engineering. True, computer science students must take some mathematics courses. But for most of them, these are a vocational requisite, not an intellectual adventure.
- *Today's students have been indulged by parents, overgraded by teachers, and are unwilling to enroll in subjects they find demanding and difficult, with mathematics an obvious example. Nor do today's students arrive avid for what professors have to offer.* I'm suspicious when instructors place the blame on their pupils. I've always thought it was a faculty's duty to impart a passion for its subject. Why isn't this happening in mathematics?

UNREQUITED LOVE

Every mathematician I've met, from small colleges to research institutes, speaks affectionately, indeed lovingly, of their calling. Peter March at Ohio State University told me that "mathematicians love mathematics, and want other people to love it too." Peter Braunfeld of the University of Illinois grew excited as he held forth about "the beauty in mathematics." Sia Wong, also at Ohio State, never ceases to be awed by a "beautifully laid out" proof.

Teachers of literature take it as a challenge to open their students to the splendors of Byron's poetry and Athol Fugard's drama. Teachers of art reveal the radiance of El Greco and the genius of Georgia O'Keeffe. In fields from music to architecture to cinema, educators seek to impart an appreciation of our enduring legacies. So why has there been so little success in stirring interest in mathematics?

The remarks of the three professors I've quoted were part of longer conversations. Here is how they continued:

> In our teaching, we offer logical arguments, beautifully laid out, and then are disturbed that our students don't appreciate our work. (Wong)

> Mathematicians love mathematics and want other people to love it too. But our problem is to get a hook into students who aren't already "into" mathematics. (March)

> The way mathematics tends to be taught obscures rather than brings out the beauty in it. Much of mathematics teaching is done as stupidly as teaching how to spell without knowing what the words mean. (Braunfeld)

Why is it that these professors, along with many of their colleagues, cannot communicate their love of mathematics and its inherent beauty? It's not as if they don't have ample opportunity. In tandem with composition, mathematics is a subject all high school students, and most undergraduates, are required to take. Indeed, departments of mathematics end up teaching well over half of the students who pass through our colleges. No other field—not history or philosophy or chemistry—has such a chance to display its wares to such a wide swath of students.

We know that few freshmen arrive at college wanting to make mathematics their major. In 2014, a bare 12,934 of the 1,672,395 who took the SAT indicated that it would be their choice. (By contrast, 66,461 were contemplating psychology, which isn't widely taught in high schools.) As it happens, most in the SAT pool will have to take at least one course in mathematics, and most do in their freshman year. I'd like to think that mathematics faculties would look upon these students as potential recruits. (Some colleges now allow options in "quantitative reasoning," which often turn out to be heavily mathematical.)

WHO'S TEACHING 101?

As has been seen, mandarins do not hesitate to hold forth on what every freshman should be taught. So it would be nice to think that they have tried out what they are advocating on their own campuses. But this isn't happening. One reason is that departments have no incentive to deploy their senior members this way. Because so many students are required to take introductory or remedial classes, the subject has outsized—if not always voluntary—enrollments even as it is attracting fewer majors. The bottom line is that mathematics departments get generous budgets for handling all those freshman sections, regardless of whether they encourage students to major in their field. Moreover, the costs of these introductory classes are low, since the teaching is done mainly by low-wage adjuncts and graduate assistants. As a result, much of the cash flow can be diverted to hiring senior faculty and giving them lighter teaching loads.

Using contingent faculty frees regular professors from tasks many of them find distasteful, if not demeaning. Stephen Montgomery-Smith at the University of Missouri openly admits that instructing freshmen and sophomores would be a waste of his talents. "It's nice having adjuncts to teach classes we don't want to touch," he explained. "If I were doing college algebra, I would get bored out of my mind." With these words, we're hearing a mandarin's definition of academic freedom: freedom from assignments they find dreary and wearying.

In 2012, the American Mathematical Society released an extensive report on undergraduate instruction. It surveyed a large sample of four-year schools, ranging from Dartmouth and UCLA, to Linfield College in Oregon and Eastern Mennonite University in Virginia. Its principal focus was the introductory courses, which most freshmen take, and thus omitted remedial sections. In one division were undergraduate tracks within larger universities, such as UCLA, which I'll call *university colleges*. The other included smaller freestanding schools, like Linfield, which I'll call *independent colleges*. Now let's see who is doing the teaching.

In introductory courses at university colleges, the average class size was a hard-to-believe forty-eight students, which of course means that about half of the classes had even more than that number. This is hardly the best setting for a subject such as mathematics, where a lot of students have questions and need timely explanations. Things aren't much better at independent colleges, where sections averaged twenty-eight students. This figure is equally disturbing, since these schools enroll only undergraduates and like to boast of their intimate atmosphere.

At university colleges, only 10 percent were taught by tenured faculty or junior staff who are in the running for promotion. (But even within this 10 percent, assistant professors do the bulk of introductory teaching.) The remaining 90 percent were given over to lecturers on short-term assignments, part-time adjuncts, or graduate assistants. This lopsided ratio underlines the extent to which university professors can exempt themselves from introductory teaching.

At independent colleges, only 42 percent of sections are taught by regular faculty. I say "only" because these schools claim to be "teaching" institutions that care about their undergraduates. Yet even these colleges hire adjuncts and other part-time staff for over half their sections. I find this not only surprising, but also dismaying, since it belies the claim that smaller colleges are firmly dedicated to undergraduate instruction.

Nor is it clear why the regular professors are excused. These colleges don't expect much by way of research, and it's unlikely they have onerous enrollments in upper-class courses. Clarence Stephens at SUNY's Potsdam campus estimates that in mathematics courses nationwide, less than 1 percent of the students are majors. What is troubling is how many independent colleges are coming to resemble universities.

WHY SCHOLARS CAN'T TEACH

Professors are expected to conduct research. This is why they have lighter teaching loads, are awarded sabbaticals, and prepare papers for conferences. Even smaller colleges, which purport to give priority to teaching, increasingly look for publications in deciding promotions. Of course, many academics feel a responsibility to advance the frontiers of knowledge. There's also the view that faculty members need to keep abreast of what's happening in their discipline. But all too much of what professors do today is writing for their academic colleagues, on highly focused topics, often accompanied by unfathomable theories.

This is especially evident in mathematics. A position paper

developed by the mathematics department at SUNY Potsdam explains why:

> The two primary activities for professors are teaching, and scholarship or research. It is generally presumed that these two activities serve to reinforce one another.
>
> In mathematics, however, they tend to adversely affect one another. The information with which a mathematics research project deals is usually inaccessible to undergraduates.

Specialization pervades every academic discipline. Even so, a professor in urban ethnography can give an idea of what she's doing to colleagues in medieval history and comparative literature. But mathematicians, increasingly, can't even explain their work to one another. Stanford's Keith Devlin, in his lectures and writing, makes mathematics come alive as few of his colleagues can. Still, he says, mathematics "has reached a stage of such abstraction that many of its frontier problems cannot be understood even by the experts." He is echoed by Ian Stewart of Warwick University, who also has a flair for enthralling lay audiences. Yet Stewart also avers, "I have never even dared to try to explain noncommunicative geometry or the cohomology of sheeves, even though both are at least as important as, say, chaos theory or fractals."

Professors making their mark in the *distribution of zeros of random paraorthogonal polynomials in the unit circle* might feel that teaching 101 should be left to lesser intellects. Given the current job market, there are plenty of adjuncts available. So it's possible that little will be gained by urging mandarins to take on introductory sections. If they don't want to and regard it as a chore, they aren't likely to approach the students in a congenial spirit. If their minds are on paraorthogonal polynomials, will they be in a mood to explain fundamentals to freshmen?

PRINCETON SETS THE PACE

For a look at what happens on the ground, I chose the mathematics department at Princeton University. It's widely believed to have one of the finest faculties in the country, setting a standard others can only hope to emulate. But for present purposes, my focus will be on its undergraduate students and how they are taught. Those choosing to major in mathematics generally average between thirty-five and forty per college class. This works out to slightly less than 3 percent of a typical Princeton year. My initial reaction was that this was not very impressive, given the department's reputation. Psychology generally attracts about seventy-five majors per year, while the department of politics enrolls upwards of a hundred. As it happens, about three-quarters of Princeton's students arrive with SAT mathematics scores that put them in the top 10 percent nationally. This suggests that in high school, they excelled in the subject. So why don't more of them choose to pursue it in college?

Let's look at what they encounter when, as freshmen or sophomores, they take an introductory mathematics course, as most of them do. Here's what I found for the spring semester of 2013. All told, a total of forty-one courses or sections were offered at the 100 and 200 levels. Yet of these forty-one, only five were taught by faculty members having a professorial rank. And four of these five were assistant professors, whose contracts will not necessarily be renewed. The one remaining class was indeed taught by a full professor. But it was an honors section, limited to students who were going to major in mathematics. The other thirty-six courses or sections—which enrolled 90 percent of the students that semester—were given over to part-time adjuncts or instructors on short-term contracts.

The fact that only one full professor teaches an introductory course tells us that, in a typical semester, twenty-four of the twenty-five senior faculty at Princeton do not deign to make contact with novice students. What are undergraduates to conclude when they find that the overwhelming majority of the regular faculty feels that their talents would be wasted on young people like themselves? Indeed, the odds are strongly against encountering an actual mathematics professor during the terms a student might be considering majoring in the subject.

Given their lofty scholarship, the mandarins should at least be able to recruit a phalanx to succeed themselves. Yet in 2013, the latest figures at this writing, only 730 American citizens completed mathematics PhDs. This is both embarrassing and revealing. After all, each year several million students are given a chance to sample mathematics in one or another college course. If the discipline is not renewing itself, it's not because it lacks importance, but rather due to the negligence and indifference of its custodians.

MAINTAINING THEIR MONOPOLY

Early in 2012, the Council of Advisors on Science and Technology submitted a report to President Barack Obama entitled *Engage to Excel*. Its main aim was to induce more young people to study STEM subjects, given increasing concern about the country's purported shortfalls in science, technology, engineering, and mathematics. In fact, there's no paucity of young people who have an early interest in these fields. The problem, the panel warned, was that "fewer than 40 percent of students who enter college intending to major in a STEM field complete a STEM degree." A study by the U.S. Department of Education found that 41 percent who begin in engineering programs either drop out or switch to another field. Subjects like physics and chemistry have even greater attrition: 46 percent. And a striking 59 percent start but don't finish in computer science. These are much higher than our high school dropout rates, but receive far less attention.

Buried in the report was what might seem an innocuous proposal. The mathematics instruction required for science, technology, and engineering, the passage said, "could be improved by having faculty from outside mathematics develop and teach mathematics courses." For example, professors in departments of statistics, or engineering, or computer science, could teach the mathematics needed in their own fields. It's not as if they didn't know their algebra.

But the president's advisors didn't reckon with the mandarins. Quite quickly, "a rumble of consternation erupted among mathematicians," according to a release from the American Mathematical Society. Tara Holm, a Cornell professor who headed the society's committee on education, called the proposal "outrageous." Her view, shared by most of her mandarin colleagues, was that only faculty with advanced degrees in mathematics itself should be permitted to teach the subject, including in introductory and vocational courses. It's redolent of doctors who seek to keep nurse practitioners from stitching a slight gash.

What Obama's advisors realized is that the "T" component of STEM refers to a congeries of technical skills. They range from operating magnetic-imaging machines to installing heating systems, and are most often taught in community colleges. Insofar as students need quantitative training, they should get it from instructors familiar with each field's needs. To give the job to pure mathematicians is a surefire recipe for failure.

The coda, of course, is that the mandarins contend that anyone not awarded their imprimatur is incapable of teaching "real" mathematics, which means only as the mandarins conceive and pursue it. That they insist on having sway over every level of instruction shows how ideology and ego can undercut efforts to address serious national needs.

VOICES

The real reason we require so much advanced math is because some kids can't do it. Colleges intentionally screen students out on the basis of their math abilities—even when such abilities are wholly irrelevant to their degree programs.

I pointed out to my college's dean that the math courses were where the most freshmen were flunking out. I was told that that's what those courses were for: to weed out the "weak" students quickly.

We are contributing to the creation of a public where all non-math pursuits are implicitly devalued, and kids by the millions are relegated to the career landfill.

What about kids who are no good at PE? Do we demand that they not receive a diploma until they master basketball as well as the best players?

VOICES

Defenders of the existing system love math because it is easy to test. There can be test-prep courses and state-wide tests and national tests and tests comparing us with other countries.

The rules are set by those who had to take multiple math courses themselves. Having done so, they believe that math is necessary for everyone and the more of it the better.

Here in Seattle, engineers and techies deride those who do not share their proclivity for abstract mathematical thinking. They refuse to acknowledge anything that does not support their narrow experience of life.

Is it worth destroying the lives of countless young people to make sure they all learn how to multiply binomials?

8

The Common Core: One Size for All

A story is told of a visitor to the French Ministry of Education, where his host declared: "It is now 11:35 a.m. I can tell you that every fourteen-year-old in France is halfway through a test on the Battle of Waterloo."

We're closer to this than you might think.

As this book goes to press, public schools in at least forty of our fifty states have scheduled tests on language and mathematics under the aegis of a transcontinental conglomerate called the Common Core State Standards. The Common Core is no longer a proposal or a pilot project. Its so-called standards are well in place and on track to be the most radical move in the annals of American education. And these standards are going to affect a lot more than what happens in our schools. Both they and the tests essential to them intend to recast our entire society.

Central to the Common Core is a universal mathematics hurdle, which will be higher than any previously set for so wide a swath of students. This by itself warrants discussion of where this grandiose enterprise came from and how it was embraced. And there's more. The Common Core's approach to both language and mathematics—science and social studies are to come later—embodies a particular conception of education, turning on the technical training and skills employers say they want and need. Nor is this curriculum the only innovation. New testing technologies will oversee how what is taught will be assessed.

UNRAVELING A MYSTERY

How the Common Core got off and running is much like a classic mystery story, and hence needs some unraveling. What can be said with certainty is that it was never discussed in legislative chambers at the state or national level, or even in their education committees. In fact, the Common Core got under way with virtually no public airing. There's been more debate about assisted suicide than whether to subject every public school student in the country to the same, unbending lesson plans.

The basic idea for a Common Core began with a little-known business group called Achieve, Inc. In 2004, the group issued a report titled *Ready or Not: Creating a High School Diploma That Counts.* Achieve's principal funders were high-tech and financial firms, like IBM, Intel, and Prudential Securities. The group's primary aim was to reconfigure high schools into training centers for the kind of workforce its sponsors wanted. While not explicitly announced, from the outset, the linchpin of the Common Core was mathematics. (Achieve, Inc. showed no interest at all in encouraging the arts, humanities, or history.) The group's priorities were apparent in a second publication released in 2008, entitled *Math Works.* Among that report's pronouncements were: "All Students Need Advanced Math," "Advanced Math Equals Career Readiness," and "Americans Need Advanced Math to Stay Globally Competitive." To bolster these assertions, Achieve, Inc. used the spurious statistic I cited in an earlier chapter: that 62 percent of all new jobs would have a need for algebra.

But Achieve had a problem. With corporate sponsors, the group was ill-positioned to make the case for a multistate campaign. So its directors struck an alliance with a not-for-profit group called the National Governors Association. This was an inspired move. At one stroke, the chief executives of all fifty states were cast as backing a vision of education based on mathematical training and technical employment. And it worked. When the heads of several university systems later lined up to support the Common Core, they said the program was produced "at the request of governors." This was not quite the truth. The governors as a group, or even as individuals, never did any requesting.

What also isn't widely known is that the National Governors Association is essentially a paper edifice. It has a meager agenda and not much to do. At its 2014 conclave, only twenty-nine of the fifty governors—barely over half—bothered to appear. In our partisan times, the meetings that matter are the Democratic Governors Association and its Republican counterpart, where like-minded leaders socialize and settle on strategies. In any event, the group, seeing a chance for some limelight, signed on with Achieve, Inc. Once pressure for the Common Core got under way, as it soon would, there is no evidence that any early or later drafts were shown to the fifty governors for comments, let alone approval.

WHEN MISSISSIPPI SURPASSES MINNESOTA

Coincidentally, during this time another low-profile body called the council of Chief State School Officers was working on a project of its own. The group has an eclectic membership. Some state education chiefs are elected, but most are career educators who began as classroom teachers, moved over to administration, and ascended the ladder. They saw themselves as professionals and used the council to coordinate with colleagues in other states.

Back in 2007, members of the council agreed they had a problem. For some years, their states had been using "exit" or "end of course" tests to decide whether to deem students "proficient" in one subject or another. But the states' tests varied widely in content and how they were graded. In 2011, for example, 91 percent of Idaho's students were scored "proficient" in mathematics, as were 85 percent in Mississippi, compared with 54 percent in

Nevada and 58 percent in Minnesota. Results like these said little about actual achievement, and their inconsistencies were embarrassing, at least to the school chiefs. So council members began talking about uniform tests and scoring methods, viewing this as a task for administrators, with no political agendas. Indeed, it's not clear how many even checked in with their governors or legislative leaders.

So part of the push for what was to become the Common Core had a simple purpose: to provide consistent statistics. States and schools and students could be ranked and compared, without getting mired in apples-versus-oranges issues. In its first incarnation, the initiative didn't involve a particular curriculum or syllabus, or even general criteria for what should be taught. Rather, as Harvard's Edward Glaeser put it, what was intended was a "common national test."

Thus all public school tenth-graders, from Savannah to Seattle, would take basically the same geometry test, which would be scored according to an agreed-upon template. As the test makers memorably put it, grades would be "norm referenced," to excise subjective judgments. The implication was that grades given for classwork were only teachers' opinions. With uniformity in place, the scores of Seattle pupils could be compared with those of their agemates in Savannah, as if they were in the same classroom and had studied identical lessons during the term.

While national testing may have been an administrative idea, schooling in America has long been regarded as a local province, notably managed by elected boards. Anything uniform in education tends to be regarded with suspicion. Thus an unusual kind of energy would be needed to install a national system. That helps to explain why the Council of Chief State School Officers joined with Achieve, Inc. and the National Governors Association to push through the Common Core. Together, they formed a powerful triumvirate: a corporate lobby, an association of administrators, and an association nominally composed of elected officials.

Ultimately, the Common Core morphed into a full set of K–12

lesson plans, spelling out what every teacher was expected to impart and every student to learn. The "exit" examinations would be retained, but now in a national guise. Twelfth-graders would be required to pass rigorous standardized tests in order to graduate from high school.

WRITING PRODUCTS AND INFORMATIONAL TEXTS

Enter another player, an educational consultant named David Coleman, who had set up an entity called Student Achievement Partners. He let it be known that his team was ready, willing, and able to get all those standards written. After all, this wasn't something Achieve, Inc., the governors, or even the school chiefs were equipped to do. A graduate of Yale, Coleman had written an honor's thesis on Edmund Spenser, and still styles himself a humanist. Yet he soon adapted to the standardized teaching and testing on which the Common Core is based. So now, instead of allegorical poetry, he speaks of "writing products," "informational texts," "automaticity," and "constructed responses."

At about that time, a final member joined the Common Core coalition: the Bill and Melinda Gates Foundation. The short version is that the Gates Foundation quickly and decisively distributed $35 million to the Common Core's sponsors, allowing them to spread their wings. Later gifts would raise the total over $200 million. This funding ensured that by 2010, the Common Core was ready to make its debut. But the first step was to familiarize teachers with the new classroom materials, which by my count totaled 1,386 new "standards." Then, spring of 2015 was set for all participating states to begin using Common Core tests for language and mathematics.

The federal government's position on the Common Core has been ambivalent from the start. On the one hand, Secretary of Education Arne Duncan seemed to disavow the Common Core's standards. "The federal government didn't write them, didn't approve them, and doesn't mandate them," he claimed. "Anyone who says otherwise is either misinformed or willfully misleading." On the other hand, Duncan's agency required that states embrace those very standards if they wanted to compete for Race to the Top funds, signaling federal approval, albeit short of a mandate.

In addition, the Department of Education put up $360 million to create multistate tests that would gauge how well students had absorbed the Common Core's lessons. In the end, the only part of Duncan's protest that seems true is that the federal government didn't write the standards. (Neither did any states or school districts.) Ultimately, Duncan became a prime cheerleader for the Common Core, in particular urging corporate CEOs to line up behind it. Here was one of his arguments:

> The child of a Marine officer who is transferred from Camp Pendleton in California to Camp Lejeune in North Carolina, will be able to make that academic transition without a hitch.

Beginning to sound a little like France?

The goals of the Common Core are more focused than George W. Bush's No Child Left Behind initiative of fifteen years earlier. (That program aimed to have every young American "proficient" in all subjects by 2014, a goal in no way attained.) The Common Core has a more precise objective, largely attuned to its employment emphasis. The phrase it recurrently uses is that all students should be made *college or career ready.* Those words need some parsing.

"College ready" is the easier part. It means that students who hope to continue their schooling will be prepared to meet the admissions hurdles that colleges set up, and then cope with courses once they arrive. Since most colleges want applicants to take ACTs or SATs, students had better have the three years of mathematics those tests demand (or four years if they want the scores demanded by Ivier schools). At no point does the Common Core question the colleges' expectations. If admissions officers want three years of advanced mathematics, the Common Core's response was to acquiesce.

PASCAL'S TRIANGLE AND PYTHAGOREAN TRIPLES

"College ready" may be suitable for students who want to proceed straight from high school to a bachelor's degree. But by no means all choose this path. A significant fraction of graduating seniors feel they've had enough of studying by the end of high school and opt to enter the workforce. (This noncollege group is disproportionately male.) For this group, the Common Core has a beguiling turn of phrase. It asserts that its lesson plans will equally benefit them, by making them "career ready." This is a serious claim and should not be made without factual backing. Yet after examining its 1,386 standards, I could find nothing at all geared to occupational preparation. Indeed, at no point do the authors or sponsors of the Common Core mention, let alone support, vocational programs. As it happens, these can be quite sophisticated. New York City, for example, has a High School of Aviation Trades. But the Common Core wants none of this.

So what might the standards mean by "career ready"? That can depend on how the word "career" is used. It need not be confined to traditional professions like law or medicine. If we look at the two-thirds of adult Americans who don't have college degrees, their work covers a wide swath and often carries serious responsibilities. Their careers can be with FedEx or UPS, as flight attendants, or supermarket managers, or technical support workers crucial to STEM fields. Perhaps even more commonly these days, workers move from job to job as, say, security guards or in construction or out on oilfields.

Still, the Common Core's stress on wholly academic subjects assumes all "careers" in the coming century will require the cerebral tools inherent in its standards. Hence it posits that in every school system and state, all students are to be held to the same curriculum and the same tests, whether or not they are bound for college, and no matter what kind of employment they plan to pursue. In a word, the Common Core views *college ready* and *career ready* as identical. Ken Wagner, New York's associate commissioner of education, was explicit on this. "College and career skills are the same," he told me. Achieve, Inc., the Common Core's original architect, spelled it out, including its own italics: "*All* students—those attending a four-year college, those planning to earn a two-year degree or get some postsecondary training, and those seeking to enter the job market right away—need to have comparable preparation in high school." On this highest-common-denominator premise, both security guards and Cal Tech aspirants will need to pass tests on Pascal's Triangle and Pythagorean Triples, as attested by these selections from the Common Core standards:

Use the properties of exponents to interpret expressions of exponential functions. For example, identify percent rate of change in functions such as $y = (1.02)^t$, $y = (0.97)^t$, $y = (1.01)^{12t}$, $y = (1.2)^{t/10}$, and classify them as representing exponential growth or decay.

Prove polynomial identities and use them to describe numerical relationships. For example, the polynomial identity $(x^2 + y^2)^2 = (x^2 - y^2)^2 + (2xy)^2$ can be used to generate Pythagorean triples.

Know and apply the Binomial Theorem for the expansion of $(x + y)^n$ in powers of x and y for a positive integer n, where x and y are any numbers, with coefficients determined for example by Pascal's Triangle.

In the past, unabashed tracking placed middle-class pupils in academic streams and assigned their working-class agemates to vocational programs. New York had high schools for baking, printing, and needle trades. There was even one in Brooklyn that proudly called itself Manual Training High School. Indeed, the aim of these high schools was to prepare a slice of society for blue-collar life. Since the 1960s, this has been construed as undemocratic. It was agreed that maintaining separate streams came close to decreeing that an entire tier of young people would be consigned to lives as bakers, printers, or manual laborers. More than undemocratic, this system meant that latent talents would remain undiscovered and undeveloped. Baking is an honorable trade, but that young woman tending an oven might be an incipient astronomer, a skill that would never be discovered if she were tracked to a nonacademic curriculum from an early age. Hence pressure arose to have all students take advanced geometry and algebra, ostensibly giving them more options if or when they graduated. The Common Core effectively enacts a high-denominator, one-size-for-all national policy.

In fact, though, the rigidity of the current mathematics sequence leads to a de facto form of tracking. Upwards of one in five of our ninth-graders now leaves high school without a diploma, and the most common academic reason is failing to fulfill mathematics requirements. Those pupils have been consigned to a lifelong track: citizens who lack the basic social credential of a high school diploma. Of course, that credential should be earned. What needs discussing is what are rational requirements and which should be mandated for everyone.

MATHEMATICS: HARVARD VS. HEATING

The flip side of conflating the college and noncollege pools means that anyone who can't surmount the summits of Pythagoras and Pascal will not be able to get the high school diploma that is now demanded for noncollege careers. As Mitchell Chester, commissioner of education in Massachusetts, one of our top-scoring states, says, "Our system isn't ready to deliver a college-ready education to all our students," adding, "I don't want students punished by not meeting that bar." In Montana, Cliff Bara, who teaches high school mathematics and science, predicts, "To say every kid in my state is going to complete Algebra 2, that's setting up some folks for disaster." In Florida's state senate, Aaron Bean says of advanced algebra for all, "mandating it is a recipe for a higher dropout rate." Anthony Carnevale sums it up: If the Common Core truly cared about career-bound students, he says, it would not inflict "Math for Harvard" on everyone, but would also offer "Math for Heating, Ventilation, and Air Conditioning."

As an alternative, Texas, which isn't using the Common Core, and Florida, which at this writing is, award several types of diplomas. Under Florida's "multiple pathways" law, enacted in 2013, students can opt for either a "scholar designation" or a "merit designation," with the latter acknowledging that not every high school graduate must have had advanced algebra. Texas has three diplomas: "distinguished," "recommended," and "minimum." The last recognizes that almost a fifth of Texas public school pupils currently aren't finishing, and represents an effort to help those students graduate.

Is having three diplomas tantamount to tracking? The larger goal is having everyone finish high school, and that can't be done without a few concessions. While some may see the Texas plan as condescending, from a more sensible perspective it's more respectful than imposing one size on everyone. Especially when that size is intended for the most demanding disciplines, and means many students will get no diploma at all. A more versatile system recognizes that individuals vary in inclinations and aptitudes, not to mention opportunities and encouragement.

Much will depend on the heights of the graduation bar. Thus far two states have had a dress rehearsal for how we can expect students to fare on the Common Core, albeit for elementary grades. In 2011, Kentucky had fully 60 percent of its pupils failing the mathematics part. Two years later, New York's mathematics rate was even worse, with 69 percent falling below the passing grade. When figures are this high, they have to cover a full social spectrum.

At this point, I'll ask to be forgiven for reiterating what I said at the outset of this book: advanced algebra is as much a stumbling block in spacious suburbs and gold coasts of sophisticated cities as it is in rural counties and inner-urban neighborhoods. Many of the letters I receive are from professional parents, whose sons' and daughters' lives have been mauled by mathematics. Nor are they misfits or whiners. These young people are accomplished in many exacting fields. Yet with its unyielding stress on trigonometry, precalculus, and advanced algebra, the Common Core will create arbitrary and intractable hurdles for students whose aptitudes lie outside of mathematics. It's much as if we required all students to play a clarinet concerto in order to graduate from high school.

What's seldom discussed is that private schools, while they must have state accreditation, are exempt from the Common Core and its standardized tests, unless they choose to adopt them. Some are repelled by its rigidity. The Hawken School, in Cleveland, Ohio, for example, takes pride in its interdisciplinary programs, such as the one that integrates economics, history, and literature. There is no way, it says, the Common Core can evaluate offerings like that. But our income distribution being what it is, alternatives like these will be available to only a fraction of those who want or need them. (Query: how many Common Core advocates have children in schools where it will be imposed?)

DISSECTING A 3-4-5 TRIANGLE

The politics behind the Common Core aren't easy to untangle. President Obama apparently backs it, but he has left the heavy lifting to Arne Duncan, his secretary of education. Still, it's not easy to find elected Democrats or committed liberals who openly support the project. While teachers' unions are not officially opposed, they have been highly critical of its high-stakes testing system.

However, there's been an active back and forth on the conservative side of the aisle. Commentator Glenn Beck sees the Common Core as "leftist indoctrination." The Republican National Committee has weighed in, calling it "an inappropriate overreach to standardize and control the education of our children." Senators Ted Cruz, Rand Paul, and Marco Rubio all oppose it, as yet another power grab from the Beltway. The libertarian Reason Foundation sees it as "crony capitalism for computer companies," especially Pearson and Microsoft, which have inside tracks for the software central to Common Core testing.

However, scholars at the Manhattan Institute and the Fordham Institute joined to laud the Common Core as promising a "more rigorous, content-rich, cohesive K–12 education." It also has a phalanx of corporate CEOs on its side, especially in firms that rely heavily on engineering. Business conservatives see the Common Core essentially as workforce training, conducted by the schools, so they don't have to do it themselves. Here, Jeb Bush speaks for his party's corporate wing. He isn't troubled by high failure rates, since they will reveal the "ugly truth" about American education and force a "painful readjustment" toward higher standards.

There's another side to Jeb Bush's position. When, as governor of Florida, he visited an Orlando high school, a student had a question for him: "What," Luana Marques asked, "are the angles on a three-four-five triangle?" She wondered if he knew, since problems like this were on a statewide test, which she and some 140,000 high school seniors had just had to take. "I don't know," her governor confessed, "one hundred twenty-five, ninety and whatever remains on one hundred eighty?"

Had a teenaged Jeb Bush muffed questions like that, he would have been denied a high school diploma under a statute he supported and signed. In fact, not three in a hundred adults could answer Ms. Marques's question. (It's not as easy as it might seem.) Supporters of the Common Core feel that parsing triangles will infuse a mental rigor that all too many young Americans are seen as lacking. Hence, its high STEM-oriented hurdles and scant attention to poetry. The suggestion that a mind can profit from perusing Emily Dickinson is barely heard in the century we are entering.

David Coleman, who now heads the College Board, wants to integrate its SAT with the Common Core, doubling the height of the barrier. His goal, he told *Education Week*, is to require of all students "the mathematics needed to pursue further study and careers in the STEM fields of science, technology, engineering, mathematics"—a stance oblivious to the fact that the vast majority of students will not end up in STEM fields. The Common Core's one-size-for-all will derail these young people at an early age, by inflicting a mathematics matrix that is not needed for all kinds of fulfilling lives.

9

Discipline Versus Discovery

In the seventeenth century, Europe had its Thirty Years' War from 1618–1648, with Catholics battling Protestants over esoteric doctrines. America's "math wars" have festered for almost that long, with a similar zeal. Despite the passing of decades, the battle lines have barely budged. So I'll start with some clippings I found at the bottom of my files. "Math becomes fun and games," a 1997 *Time* article observed. But then it wondered, "Are the kids really learning anything?" *Newsweek*, also that year, let the two opposing sides in the math wars show how they would set up lessons:

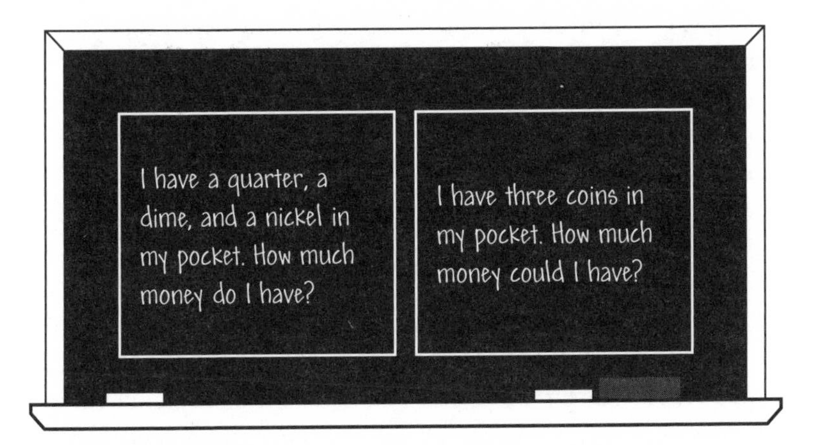

The problems may seem simple. But they have powerful underpinnings. In the first, which depicts what I'll be calling the *Discipline* school, the message is clear. There is only one correct response, and it's what you need to get. The second, often called the *Discovery* approach, doesn't have a preset answer pupils are expected to find. Just to start, there's its ambiguous word "could." We're not told whether to look for the largest possible total, or all feasible variations. That, presumably, is left to the students, perhaps in a collaborative discussion. And there's room for more thought and creativity: Are the coins American? Could I have bills in my pocket also? Both schools claim that they prepare youngsters for the real world. One says we need accurate results, which calls for mastering basic rules. The other says that in much of life there isn't one single solution.

We've seen a similar debate over how to teach reading. On one side is the "phonics" approach, which stresses sounding out words. Its first premise is that rules and regularities must be memorized, which demands disciplined drill. Fervent supporters of phonics tend to be conservative in their social and moral outlooks. The other side is usually called the "whole language" method, which stresses recognizing words and discerning their contexts. Instead of drill, it seeks to impart a love of reading, by putting books in children's hands as early as possible. Proponents of this approach are more likely to hold liberal views on broader issues. (Of course, most Discovery proponents grant the need for basic memorization. Instilling a reflexive response to 9 times 7 requires repetitious practice.)

IDEOLOGIES AND INTERESTS

Having been through the mill ourselves, most adults feel entitled to expound on education. Mathematics is a frequent focus. It was the hardest class for most of us, and is notable for low and failing grades. Its difficulty, combined with its being required of all students, fuels much of the Discipline/Discovery debate. Those on the Discipline side say there's no honest way it can be made easy. Here's a mantra recited by sixth-graders in New York's Bronx Preparatory Charter School. When their teacher asks, "What is math?" the students respond:

This is math. I don't have to like it to pass it.
I don't have to enjoy it to learn it.
I don't have to love it to understand it.
But I must, and I will, master it.

"Mathematics is hard," says John Derbyshire, a conservative columnist and self-styled "pop-math author," who adds, "probably no large number of people will ever be much good at it, or like it much." If he is right, it's in order to think about how to teach a subject where most students will never be "much good." I'm sure that some youngsters may find a three-coin problem perplexing. Even so, they should be made to stick with it, since handling that kind of calculation is part of preparation for adulthood. Whether this also holds for the advanced mathematics now required in high schools is, of course, a recurrent theme of this book.

From one standpoint, the battle pits two phalanxes of professors against each other. On one side are research mathematicians at elite universities, a group I earlier styled as mandarins. They may be joined by academics at lower-tier institutions who elect to identify with them. I've used the Discipline rubric for their position because they stress the tenacity needed to penetrate their subject. Moreover, they seek to preserve and promote mathematics as an intellectual pursuit and a sphere of study. So why are these mandarins and their acolytes so concerned with what occurs in the third grade? After all, our coins-in-pocket problem doesn't call for advanced algebra.

Or perhaps it does. Here, another historical vignette can help. Back in 1999, Bill Clinton's secretary of education gave his imprimatur to a report that cited several Discovery programs as "exemplary" or "promising." A firestorm soon erupted. It took the form of an open letter, ultimately signed by more than two hundred mathematics professors. While most were from elite universities like Harvard, Stanford, and MIT, space was given to signers from institutions like Framingham State College and the University of Northern Florida.

The letter began by complaining that the "panel that made the final decisions did not include active research mathematicians." It then explained why scholars, as the signers largely were, had strong opinions about what happens even in the lowest grades. In their view, even the teaching of addition and subtraction should be part of a seamless preparation for the mathematics that will be encountered not only in high school but all the way up to post-graduate seminars. In a word, even first-graders should be treated as incipient mathematics majors, if not eventual aspirants for advanced degrees.

This view was made more explicit in a separate pronouncement prepared by a group of professors from Rochester, Johns Hopkins, and the University of Southern California, sent out in the early days of the Internet. The manifesto explained how and why they wanted everyone to be agile at long division. On its face, this is something we should all know how to do. I too would hope everyone can divide 24 into 300, preferably in their heads. Or divide 67 into 13,684 with pencil and paper, and without a calculator. But the mandarins have much more in mind. For them, 24 into 300 is part of a long-range plan:

> Long division is a pre-skill that all students must master to automaticity for algebra (polynomial long division), pre-calculus (finding roots and asymptotes) and calculus (integration of rational functions and Laplace transforms).

So according to the mandarins, students need to be prepared for concepts like *Laplace transforms*, which are usually encountered in courses geared to mathematics majors.

PRECISION AND PERSEVERANCE

The Discipline/Discovery debate involves conflicting views of human nature—or youthful nature—joined to notions about what it takes to sustain a successful society. The Discipline school exists in major respects thanks to William McGuffey, the nineteenth-century textbook author, who felt that young people should learn that academic regimens will be firmly enforced. Karsten Stueber of the College of the Holy Cross put it succinctly: "Skills must be mastered, and a certain amount of drudgery must be endured." There's a lot in life we don't like, and mathematics is a good way to teach young people to get a grip on themselves.

A further McGuffey prescription is that education should shape character. Hence, its stress on inner strength, postponing gratifications, and acquiescing to authority. Mathematics, unlike other disciplines, is less susceptible to fads, fashions, and fleeting opinions. In the view of Simon Singh, an expert on Fermat's Last Theorem: "More than any other discipline, mathematics is a subject that is not subjective." Discussion may be suitable for social studies. But geometry entails a corpus of skills young people have to learn. Nor should students be expected to come up with innovative mathematical ideas. If they think they have, they are likely to have missed the point.

Also underlying the Discipline approach is this country's puritanical penchant, which never goes away, but reemerges in new forms. While those on the Discipline side don't say it explicitly, more than a few of them are suspicious of lessons that pupils might actually enjoy. At the college level, they are scornful of professors they regard as entertainers, or are seen to curry student favor. In lower grades, they warn of teachers who make their subjects easy to gain rows of happy faces. Down deep, the Discipline school believes that "no pain, no gain" applies as much to mental strain as muscular exertion. These views were encapsulated in many of the responses to my *New York Times* article. I've reproduced an abbreviated sampling on the following pages:

VOICES

We should require algebra for the same reason that military basic training requires push-ups. Even if you never use it in battle (life), it will make you stronger (smarter).

Students able to persevere and solve problems are more likely to be successful than students who throw in the towel at the first sign of difficulty.

Math, like many things in life, demands precision. There is a right answer and a wrong answer. No credit for being "close."

Should we eliminate everything that is hard? Education should build character, it shouldn't be a joyride.

VOICES

I'd be worried about being treated by a doctor who didn't have the ability and perseverance to learn basic calculus.

I recall hours in my room, toiling and grinding away. This taught me much more than mathematics: that patience, persistence, and practice pay off. A life lesson.

Rigorous discipline is essential in mathematics, and a lot of students may not like the rigor. Sure a 95-mile-per-hour baseball is hard to hit. But you pitch at 35-miles-per-hour, you won't get to the major leagues.

What first struck me about these comments was the severity of their tone, coming close to being bellicose. It's as if the writers, in having been made to master mathematics themselves, had survived a physical ordeal, which girded them for success in an arduous world. Indeed, there's an air of superiority, confirming their status as moral models. All the allusions to character and discipline, persistence and rigor, precision and patience are presented as traits they imbued in themselves.

Sadly, there are no mentions of the beauty of mathematics, its intellectual provenance, or its place in the natural universe. Their most striking references are to push-ups, baseball, and boxing ("throw in the towel"). For at least a few of them, mathematics is a metaphor for national supremacy, economic preeminence, and a resolute citizenry.

Also revealing is the fact that Discipline advocates like to see classrooms configured in a conventional grid, with pupils sitting in rows of desks and the teacher at the front of the room. During most of each hour, students have their textbooks at designated pages, while the teacher guides them through exercises, usually working out problems on a board. From time to time, some pupils may be sent up to the board. But for the most part, they remain at their desks, presumably attentive to the teacher.

Mathematics, perhaps more than other subjects, favors pupils who give precisely the answers their teachers want. Perhaps for this reason, there's less inclination to indulge students who don't keep up. So Cs and Ds and Fs are more usual in mathematics than in other subjects. The notion of a single correct response is also considered good preparation for the ACT, the SAT, and the Common Core "assessments," which many states will use to decide who will receive high school diplomas.

JOHN DEWEY ENDURES

On the Discovery side are faculties in colleges of education, which produce most of our K–12 mathematics teachers. Few of these professors are known beyond their campuses, and sometimes not widely even there. Yet from Alabama to Alaska, their students staff the great majority of the nation's classrooms.

Most professors of education still carry the torch of John Dewey, which they seek to pass on to their own pupils, who will disperse to teach arithmetic and mathematics from kindergarten through the twelfth grade. These professors hope to structure learning at all levels in a way that students find interesting and engaging, plus as enjoyable as possible. At the high school level, they are particularly disturbed by the high failure rates in mathematics classes, which they ascribe more to misconceived lessons than to recalcitrant children. At least until recently, professors of education have had the greatest influence over how subjects such as mathematics are taught in school.

The Discovery approach espoused by education professors begins by urging young people to analyze problems and come up with solutions on their own. It has also been called *constructivist*, since it encourages students to build their own strategies and skills; or *interactive*, to emphasize give-and-take in the classroom. Hence, one of its premises is that a "mathematics classroom is a community of learners rather than a collection of isolated individuals." (This statement and others I'll be drawing on are from *A Teacher-to-Teacher Guide* put out by Interactive Mathematics Program, a research and advocacy consortium.)

In a typical community, a class of twenty-five students might be broken into five groups, each with its own table. The teacher strolls around the room, stopping to confer at each table, occasionally addressing the class as a whole. Pupils working on a problem can ask the teacher for assistance. But this should be allowed only when they have "questions which could not be answered by the combined efforts of the group."

In small groups, students will be more likely to take risks than in a larger class. With only five at the table, shier members will be more prone to participate, especially when the teacher stresses that everyone should be involved. So as they "hear each other's approaches, students take the responsibility for each other's work habits and classroom behavior." A further finding is that distributing students randomly works best for everyone's learning. It's been found that pupils who arrive at solutions rapidly will also gain, since "one of the best ways for students to improve their reasoning is to explain or justify their solutions to others." (This is also a strong argument against segregating ostensibly "gifted" pupils in separate classes and schools.)

Discovery advocates believe that "students learn best when they construct their own knowledge." Moreover, being encouraged to devise their own methods encourages them to exercise their own intellects, as well as imagination and ingenuity. Here's how Discovery might work with a problem involving long division.

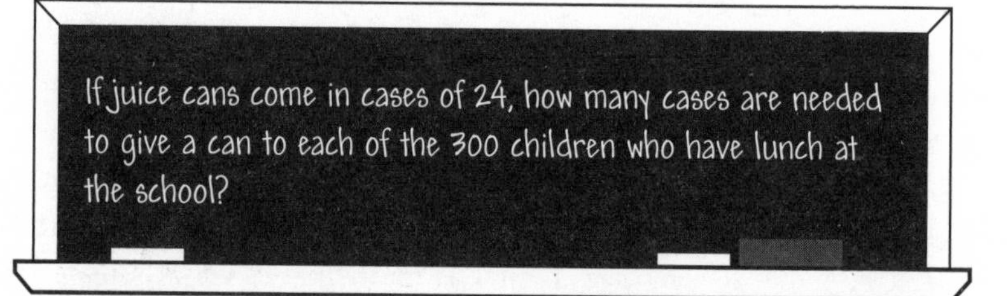

In a Discipline setting, pupils are taught to embark on the division of 24 into 300, which yields first a 1, then a 2, and then, after a decimal point, a 5. They can present their answer as 12.5, or 12½, or perhaps 12-remainder-5. They may then say that the .5 or ½ represents half a case. Therefore, they'll note that the 288 cans in 12 full cases will be 12 cans short of what's needed for 300 students. The formally correct answer is 13 cases.

How might a Discovery class deal with the juice problem? Its focus will be not simply on securing a single correct answer. The exercise should also prompt them to wonder *why* they are being asked to come up with a response. After all, numbers are not only abstract symbols, but tools humans have devised to find solutions to practical problems. I've taken the liberty of imagining the reports from five classroom tables.

Table A: We decided to just get 12 cases. We know that this will leave us 12 cans short. But on most days, at least a dozen of the 300 students will be absent. And of those present, some don't like the juice and won't drink it.

Table B: We'd order 12 cases, which gives us 288 cans. If we find that more than 288 kids are going to be there, we can pour from some of the cans into paper cups, and that will provide enough to go around. No one will really lose more than one swallow.

Table C: If we place an order for, say, 25 cases, that will be 600 cans or enough for two lunches. Then we won't have to go through this long division every time.

Table D: We discussed various approaches. But we decided to answer the question directly, which is that we'd need 13 cases if all 300 kids showed up, even if that means some of the cans won't be used. Is that the "correct" answer you wanted?

Table E: After doing the division, we found that another issue arose. It's that a lot of kids don't finish all the juice in the can they're given. (A lot of half-filled cans are left on the tables all the time.) Some other kids might really like more; but they are not always seated near people who might give them some of theirs. So instead of cans-and-straws, we suggest placing juice pitchers and paper cups on each table to see what happens. It may even end up cutting costs.

If classes are conducted this way, Discovery proponents say young people will become better problem solvers. Learning to share ideas is good preparation for the adult world. Indeed, it's what mathematicians themselves do, when they discuss problems over coffee, debate at conferences, and scribble over each other's work. A Discovery syllabus reminds us that "genuine mathematical work, whether done by mathematicians, plumbers, engineers, or dental hygienists, typically involves collaboration and communication." Its implication is that the stress on individualism and competition of the Discipline approach ignores this communal component in creating and transmitting knowledge.

The Discovery ideology also sees each pupil as an inquiring intellect, an imaginative creator, an incipient artist. Young people should be urged to evolve ideas of their own and be prepared to express them. So a teacher should be not "a sage on the stage," but "a guide at the side." This doesn't mean that fifth-graders are expected to come up with original theorems. But they can—and have—created insightful solutions. Young people who are so encouraged may not absorb as much formal mathematics as the Discipline school feels they should. But they are learning to use their minds and ingenuity, which some people feel is the purpose of education.

Needless to say, Discipline supporters don't agree. For one thing, they may wonder if some of the students are freeloaders, who can't or won't do the long division. More crucially, they may simply ask how much actual mathematics is being learned. This is the view of Al Cuoco, who taught algebra for many years at Massachusetts' Woburn High School:

> I am worried that people ten years from now will say, "I'm good at mathematics and I like it," but the "mathematics" they are good at and like will be quite remote from the mathematics used by scientists and mathematicians.

Literally, he's right. But his remark rises from a premise that everyone who enters our schools should be compelled to master "the mathematics used by scientists and mathematicians."

WHAT'S A TEAM?

In an earlier chapter, I mentioned the International Mathematical Olympiad in examining whether mathematics needs an atmosphere of free expression in order to thrive. (The answer: it doesn't, as seen by its flourishing in Iran and North Korea.) I want to revisit this annual event, but now for a different reason. Here, I'm not specifically interested in the winners, but how the competition is conducted.

As was noted, the 2013 gathering in Santa Clara, Colombia, was set up to have teams with six members from each of the ninety-seven competing countries. (It turned out that Luxembourg sent only two, Montenegro had four, and Liechtenstein made do with a single participant.) So in the end, 572 young men and women took part in the events. I will use this occasion to ask what we mean—or want to mean—when we refer to a group of people as a *team*.

In the most basic sense, a team is a group of individuals playing under a common banner, like the U.S. gymnastics team or the cross-country squad at Iowa State University. That's the easy part. Everyone on the roster is a member. But *team* can have more exacting connotations, as happens when we begin to talk about *teamwork*. Of course, we like to feel that the gymnasts and runners cheer their teammates on. Still, most of the scoring in such sports comes from the success of their individual members. But in other sports, like soccer and basketball, success depends on working well together. (Baseball and football occupy more of a middle ground.) Now to mathematics.

Despite the fact that the Olympiad records scores by teams, the event is still a competition among *individuals*, more akin to gymnastics than to basketball. That is to say, all 527 contestants are expected to solve mathematical problems entirely on their own. Thus each one of the 527 sits at his or her own table, addressing identical problems. At the final bell, they hand their personal answer sheets to the judges. The scores for each country's entrants are totaled, which determines the final national ranking. Hence in 2013, the United States placed third with 190 points, which was the sum of its entrants' 35, 35, 35, 31, 29, and 25. China again came first with 208, as it has in eighteen of the twenty-nine Olympiads. Other totals were 184 for South Korea and 168 for Iran.

For my part, I find it hard to see how a competition based on the separate scores of individuals can be said to involve a *team* in any serious way. If we look at the United States, its six members were as much pitted against one another as facing entrants from other countries. This of course introduces a double edge: while entrants want their country to come in first, each member might also hope to be the highest within his or her own six.

Perhaps I shouldn't be unsettled by the Olympiad's system. After all, in most high schools and colleges across the world, mathematics is treated as a personal pursuit. As a student, you are awarded your own A or C or D, based on your individual work, as you earn your 660 on the SAT. Until the moment tests start, students are usually allowed to share ideas with one another, as they do in law school study groups. However, the instant an examination begins, all must be absolutely *incommunicado*. Making contact with anyone else is deemed an academic felony. But could so much stress on personal performance in fact be self-defeating?

I think so, which is why I was disappointed with the Olympiad. I am sure team members socialized while en route to Santa Clara, even up to the moment they went to their testing seats. So why not let them continue this collaboration on the playing field, as happens with basketball and soccer? The basic idea behind a team is that its members have varied strengths and styles, which combine as they confront a common challenge. This is why I find myself feeling that it would have made more sense to seat each country's members around a table of their own. There they could mull over the problems and settle on the answers they would present. The judges would then score the 97 national submissions, instead of the 527 they recorded in 2013.

True, many mathematical discoveries are associated with individuals. We assume that Bertrand Russell worked out his paradox himself, just as Bernhard Riemann arrived at his zeta function on his own. Yet even in this intellectual stratosphere, it often emerges that two minds do better work than one. That's why we have the Calabi-Yau manifold, the Hausdorff-Besicovitch dimension, and the Kuder-Richardson formula. Indeed, in mathematical journals, most articles now have two or more authors. Or observe a group of professors arguing as they ponder an equation on a board. "Most problems require the insights of several mathematicians in order to be solved," Sylvia Nasar and David Gruber point out. "Mathematics, more than many other fields, depends on collaboration."

10

Teaching, Tracking, Testing

Whenever students are felt to be performing poorly, the first target tends to be their teachers. Among the charges are instructional incompetence, resistance to innovation, or focusing on faculty privileges instead of academic excellence. Teachers' affinity for energetic unions is seen as worsening the problem. We are told that their contracts provide lavish benefits and thwart accountability, while their political activism stymies legislative action. Other attacks are directed at their training, with colleges of education seen as promoting half-baked fads and feel-good diversions.

No other public profession receives such scathing scrutiny. Firefighters and forest rangers apparently do no wrong, and it's hard to recall outcries over highway engineers or public defenders. Yet there's a presumption that a critical mass of teachers are incorrigibly incompetent. No less revealing is how few of teachers' critics have spent time in a classroom, trying to instill a love of learning in preoccupied pupils.

ASSESSING TEACHERS, RATING STUDENTS

Much is also heard about the capacities of those who choose to join the profession. An oft-heard accusation is that they come from the bottom of the mental barrel. According to a report called *Tough Choices or Tough Times*, from the National Center on Education and the Economy,

> We recruit a disproportionate share of our teachers from among the less able of the high school students who go on to college.

I propose we reflect for a moment about what we're hearing here. The group being daubed as "less able" consists of *high school students*. We are being taken back to when teenagers take the SAT, at which time they are also asked what they then think might be their college major. Not surprisingly, some say education, although fewer than in the past. (In 2014, only four percent said they wanted to be K–12 teachers, compared with eight percent as recently as 2005.) In an accompanying report, the College Board compares contemplated majors with the SAT scores the students received. It's altogether true that those who check teaching tend to rank lower than classmates who pick some other majors. Those who select teaching average 966 on the combined language and mathematics tests, compared with 1049 for prospective history majors and 1107 for engineers.

But there's a reason the SAT scores are used. It's that they're the only gauges of "ability" the critics have. For better or for worse, there aren't such precise assessments once individuals reach their adult years. I'd love to see digits gauging the proficiency of bankers and surgeons; or, for that matter, mathematics professors. Or, to return to SAT results, are students who plan to study history somehow "less able" than engineers? It seems a stretch to say that test results at the age of seventeen portend occupational performances a decade or more later.

Another common observation is that talented women in the past took to teaching because they had fewer options. Today, we are told, the most competent members of their gender become scientists and attorneys, leaving classrooms to those who can't make it in more demanding professions. In this view, women now on Wall Street would be more effective than the teachers we currently have. It's an intriguing hypothesis, but I've never seen it tested. A related assumption is that if teachers' salaries were substantially increased, many of those bankers would opt for education. (But considering what can be made in fields like law and medicine and finance, it's doubtful that even the best-paid teachers would come close.) Here too some evidence would be helpful. Is a talent for leveraged buyouts applicable to bringing out the best in second-graders?

Mathematics teachers face special challenges. Most of them confront students who find the subject difficult and more than a few teachers fail. It's hard to think of other disciplines that are so demanding. Not surprisingly, some people find themselves wondering whether aptitudes for mathematics are universally distributed. Nor are teachers immune from such musings. Here's a statement that was placed before mathematics teachers in an international survey. They were asked if they agreed or disagreed:

> Some students have a natural talent for mathematics and others do not.

In fact, most agreed. In Russia, 93 percent did, as did 95 percent in the Czech Republic. Danish and Italian teachers were least likely to concur, but 65 percent and 74 percent of them still did. Of the American teachers who were sampled, fully 82 percent said that they agreed. While these are only a dozen words, they encapsulate an assumption with huge implications. As a mind-set, it explains a great deal about what's happening in our classrooms.

To speak of "talent" in connection with mathematics places it alongside pursuits like music and acting, even leadership and athletics. For that matter, students are rarely spoken of as having a "talent" for history or geology. Faculties in other fields feel that, with application and encouragement, most pupils are capable of estimable work. And there's the "natural" part. This survey tells us that four in five of our mathematics teachers believe that success in their subject calls for special genetic endowments. How many schools warn parents that their children may face this preconception?

For my own part, I believe that we all arrive on this planet with nascent intelligence and imagination, plus dollops of creativity and ingenuity. Indeed, everyone has a potential to excel in some endeavor. Does this mean that with study and support, everyone can attain 750 on the SAT, as only three percent now do? No, that's not how human beings—and human populations—are constituted. (Nor, as we've seen, are the ACT or the SAT even reliable measures of mathematical knowledge and understanding.)

Howard Gardner said it best a generation ago in his classic *Frames of Mind*. We misrepresent reality and inflict a lot of damage if we say that some people are "more intelligent" than others. Gardner posited there are "multiple intelligences." It doesn't matter whether we call them talents or aptitudes, abilities or gifts. It's instructive that we don't ordinarily describe accomplished poets or sculptors or chefs as "intelligent," if only because they draw on more than their minds. This also holds for mathematicians whose work is exceptionally original. What they've achieved chiefly arises from intuition and inspiration that is more than merely mental.

In fact, here it is conservatives who adopt an idealist stance. They believe that mathematics, properly construed, will always be demanding. Still, they add that if young people are made to buckle down, their scores will soar. As evidence, these advocates point to mathematics programs—usually in well-funded charter schools—which spur low-income and disenfranchised children to excel. Thus they contend that once a rigorous regimen is in place, no child need fail in our schools. The real barrier, as George W. Bush put it, is a "soft bigotry of low expectations." They counter that a rigorous mathematics regimen will give every child a chance to show his or her mettle. Whether they are willing to provide adequate resources for such a venture is, of course, another story.

"Unfortunately, little is known about what effective teachers do to generate greater gains in student learning": this confession came from a commission named by George W. Bush, which was charged with improving mathematics education across the nation. If so august a body, laden with academic luminaries, can't answer so crucial an educational question, then the nation has a rough road ahead. (It should be added the commissioners' method was to listen to testimony and review reams of research. There were no indications that they sought out and then observed successful classes in St. Paul or San Antonio.)

Perhaps in hopes of some hints, the panel sampled several hundred algebra teachers, asking them what was their most daunting challenge. Here's the box most of them checked: *working with unmotivated students.* Notice: they weren't complaining that their pupils were inadequately *prepared* for the subject. Rather, their lament was that their students didn't arrive with a desire to imbibe what their instructors would be teaching.

My own view is that job one for teachers, at all grades and levels, is to excite their pupils, especially those who arrive not caring much about the subject. In mathematics, there is little reason to presume that most pupils show up eager to explore azimuths and asymptotes. Yet there are teachers who can and do surmount this lack of interest. I'd like to think that all of us had at least one or two ourselves, and I've observed many while working on this book. Good teachers take their text from Las Vegas: you make the most of the hand you've been dealt. They don't blame their pupils, just as you wouldn't want to use a doctor who always faults his patients. "Are they any good?" my Manhattan friends sometimes ask me, alluding to my municipal college students in Queens. My reply is simple: "We make them good."

DO WE TRACK?

Those who like the idea call it "ability grouping." Those opposed prefer the term "tracking." The first view says it gives all young people the best chance to show what they can do. The second claims it's undemocratic and unfair, consigning some students to tiers where they will probably stay. Dividing pupils by ability—which likely ends up dividing them by their background—is more common in mathematics than any other subject. Few schools have fast and slow sections in biology or social studies. So here we have yet another consequence of forcing advanced mathematics on everyone: it segregates students. Ostensibly, the apartheid is solely for academic reasons, but the human fallout can be enduring.

Michigan offers the best case study I've come across. Researchers at Michigan State University examined the transcripts of some 14,000 high school seniors in seventeen diverse districts. The research was impelled by a statewide ruling that required all students to take four years of high school mathematics. The study's first finding was that this ruling didn't do the trick. In fact, only three of the districts—all affluent suburbs—had close to full compliance. Among the other fourteen districts, only five had most of their students meeting the mandate. In one, it was an embarrassing eight percent. These students were not measuring up for a familiar reason: they had failed geometry or algebra, and hence didn't have those subjects on their transcripts.

The study concluded that while there wasn't formal tracking *within* individual schools, the state's districts had become a de facto tracking system. Some districts took it for granted that most of their students would take the full mathematics programs, and that expectation tended to be fulfilled. But most districts assumed that their schools and students would settle for less, and that supposition turned into a fact. The gist of Michigan's experience: "Rigid curricular programs that neatly divide students have largely dissolved. This does not mean, however, that schools do not track students. Most do."

SINGAPORE, KOREA, AND THE UNITED STATES

By now, it's widely known that, taken as a group, young Americans aren't doing well on worldwide mathematics tests, where all participants confront the same questions. The two principal studies—the International Mathematics and Science Study (TIMSS) and the Program for International Student Assessment (PISA)—look beyond scores to compile information on students' feelings about the subject, the quality of teaching, and how instruction is structured. While they use different scoring systems, the TIMSS and PISA tests are similar, as are the national rankings that result.

The table on the next page gives the PISA scores for 2012, which show the United States in thirty-second place. The full field was fifty-nine countries, but I've listed only the top thirty-five.

Here are further findings, from other TIMSS and PISA surveys.

- Eighth-graders were asked if their friends gave importance to "doing well in mathematics." In Israel, 55 percent said that was so, as did 47 percent in Singapore and 45 percent in Thailand. However, only 15 percent of American pupils said those they knew admired excelling in mathematics.
- Ninety-seven percent of Korean eighth-graders and almost as many in Japan study mathematics in classes with over thirty pupils. In the United States, only 16 percent are in classes that large. Yet Asian students do markedly better. Other countries also make less use of technology, like overhead projectors, electronic boards, or desktop computers.
- Almost a third of American eighth-grade teachers allot in-class time for doing homework. In Japan, only two percent do. Israel, Hungary, and Germany expect all homework to be done at home. American teachers seem to fear that if it is left for later, it may not get done at all.

Mathematics Scores 2012

Program for International Student Assessment (PISA)

Rank. Country	Score
1. Singapore	573
2. Korea	554
3. Japan	536
4. Liechtenstein	535
5. Switzerland	533
6. Netherlands	523
7. Estonia	521
8. Finland	519
9. Canada	518
10. Poland	518
11. Belgium	515
12. Germany	514
13. Vietnam	511
14. Austria	506
15. Australia	504
16. Ireland	501
16. Slovenia	501
18. Denmark	500
18. New Zealand	500
20. Czech Republic	499
21. France	495
22. United Kingdom	494
23. Iceland	493
24. Latvia	491
25. Luxembourg	490
26. Norway	489
27. Portugal	87
28. Italy	485
29. Spain	484
30. Russian Federation	482
30. Slovak Republic	482
32. UNITED STATES	481
33. Lithuania	479
34. Sweden	478
35. Hungary	477

One lesson from these international comparisons is that young people elsewhere seem more willing to apply themselves to their mathematics assignments. They apparently perform ably in larger classes, do their homework at home, and admire peers who excel. If they get low grades, they assume it must be their own fault, and the way to do better is by trying harder. Were Dutch or Swiss schools to replace mathematics with, let's say, studying crossword puzzles or ancient hieroglyphics, we can assume that students in those countries would rise to these challenges and lead the world in acrostics and cuneiform.

To discover that one's country ranks thirty-second is not exactly heartening in a nation so occupied with winning. Hence, we see a welter of studies and reports, replete with policies and proposals, aimed at elevating attainment in mathematics and allied fields. Here, another point is important. The PISA and TIMSS tests are given to full-scale samples of students, from all backgrounds and income levels. So if the United States is to do better, improvements will have to occur in every sector of the society.

If we want to match the scores of the top countries, we would do well to acknowledge what this will entail. A more rigorous curriculum is the smallest part of what will be needed. More essential is instilling a respect, on the part of pupils and parents, for academic authority. This means bolstering the belief that schoolwork is critically important, indeed eclipses all other interests and activities. (High schools in high-scoring countries have nothing like the United States' athletic programs, for example.) It remains to wonder how many American teenagers and their parents are willing to make so assiduous an academic commitment.

Nor is it only a matter of attitudes. Compared with most other developed countries, the United States has more of its people subsisting on marginal incomes, often with little prospect of escaping their condition. While race is a major reason why these gulfs and gaps persist, a lot of white Americans are also finding themselves falling behind. Conditions like these are bound to have an effect on scholastic attainment. A nation that leads much of the world in inequality cannot expect to have stellar mathematics scores across the board.

But should we enroll in this competition? Before doing so, it would be well to ask what winning scores show. If the concern is employment skills, American workers turn out just as high-quality vehicles in Michigan and Mississippi as their counterparts do in Seoul and Osaka. That fine-tuned BMW your neighbor drives was most likely assembled in South Carolina, where enrollments in algebra are among the lowest in the nation. What makes for a competitive edge isn't a workforce with high mathematics scores.

LESSONS FROM ABROAD?

Let's see what we might learn from two of the highest global scorers, Japan and Korea. Essentially, they take two quite different routes. A 2010 OECD report stressed the central feature in Japan's approach to education. Its main finding was that "there is no tracking in Japanese schools, classes are heterogeneous, and no student is held back or promoted on account of ability." Nor need such variations lead to dumbing down: "The expected outcomes are not set at the lowest common denominator, but at the top of the range of possible outcomes." Teachers work individually and consulting with one another "to make sure that all students keep up with the curriculum." Apparently, it works in mathematics. Japan's students averaged 536 on the most recent PISA test, against 481 for the United States, where de facto tracking remains the rule. Also, unlike the United States, "there are no special classes for the gifted, nor are students pushed ahead if they are perceived to be exceptionally able."

As it happens, Japanese classes typically have from thirty-five to forty-five students, considerably more than in other countries, nor are teachers given aides or assistants. Instead, "the system is set up so that high-achieving students can help lower-achieving students within a group." In fact, this helps both sides. Pupils who are more advanced do even better when they are asked to explain *why* they know what they know.

Korea's high schools do have two streams: one aiming for college preparation and the other a less academic program. But students are allowed to choose which one they want. While social class may intrude here, specified grades or scores are not needed to embark on the college route. But there's something more striking about the nation's mathematics success. The source I found most enlightening was a 2011 research report in the *Korean Journal of Pediatrics,* based on a study of 3,370 pupils in grades five through twelve. The chief finding: "Korean adolescents have severe nighttime sleep deprivation and daytime sleepiness because of their competitive educational environment."

And there's a single reason. Virtually all Korean teenagers pay for personal coaching or attend test-prep classes that commence directly after school. These tend to go on until 9:00 or 10:00 p.m., after which students put in several hours of homework. Eleventh- and twelfth-graders reported they got between five and five and a half hours of sleep on a typical night. One result is endemic nodding off during daytime classes. Girls, in particular, prepare for this by strapping tiny pillows to their wrists to catch naps at their desks. The long hours of tutoring seem to pay off on the PISA tests. However, the attendant sleep deprivation leads to "increased emotional lability and irritability both at school and at home," as well as "inattentiveness, depression, and elevated suicide rates."

The phrase *burn out* may be a cliché, but it's now a reality for more teenagers than ever in human history. Even sadder, each year sees the test regimen beginning at earlier ages and grades. The United States is already well onto this path. A prominent rationale for the Common Core is that it will produce test results that can be compared with those of other nations. For a century, this country led the world in defining what education meant. Now it is scrambling to jump over Iceland and Estonia. Americans should ask themselves whether this is the path they want their children—and their country—to take.

TESTING VS. TEACHING

All teachers—and I include myself here—grant that tests have an important role in education. And not just for grading students. Since our job is to impart information, argumentation, and analysis, we want to know what—and how much—is being absorbed. Classroom responses help, but they aren't enough. The aim of carefully constructed tests is to improve instruction. Indeed, teachers become adept at designing them. They know what they've wanted to cover and want to learn how effective they've been.

But the current commotion over testing isn't about tests that teachers construct. It's about so-called standardized tests, which are turned out by nonprofit empires (SAT, Common Core) and for-profit corporate conglomerates (McGraw-Hill, Pearson). A not small reason they are being used is that a lot of important people don't trust teachers or the evaluations they give, and thus want an outside measure of what students really know.

By now, we all know what *standardized* means, but indulge me a minute to cite its chief features. (Or skip this paragraph.) First, everyone gets the same, or at least a comparable, test. Second, questions have only one correct answer, which students must select from several options, or enter words or numbers a computer can read. Third, the tests purport to be objective, because they are graded by electronic matrixes, or by humans following a rigid template. Fourth, they almost always have severe time limits, like seventy-five minutes to answer sixty questions.

Of all subjects, mathematics is best suited to the standardized format. Or it is, if what's wanted is choosing predetermined answers under a fast-ticking clock. But current sponsors want more. Founders and supporters of the Common Core claim that its techno-testing will reward *critical thinking, logical reasoning*, and *higher order skills.* At least some observers are wondering how such complex processes can be revealed via multiple choices or filling in small spaces.

Alternatives exist. Even now, there are schools that emphasize research projects and cross-disciplinary portfolios, supplemented by conferences where pupils can explain their work. Of course, this is costly and time consuming, and—some might say—susceptible to biased evaluations. My own question is how far techniques like these might be used with mathematics. Is it feasible to ask for creativity in geometry and originality with algebra? I think so. On the next page, I've reproduced a question that I've used in my Numeracy 110 class. It's based on one from a TIMSS test that was one of the hardest ever deployed. Worldwide, only 10 percent of pupils completed it successfully, with 4 percent of Americans getting it right. Yet what's interesting is not the actual answer—which is a two-digit number—but how to obtain it. A quick glance at the problem makes it seem fairly simple. It has no complex equations or procedures for proofs. So shouldn't elementary geometry enable us to find the length of a string? Not necessarily. What's needed here is a readiness to think outside the box. Or as the impresario Sergei Diaghilev challenged artist Jean Cocteau, "Astonish me!" ("*Étonne-moi!*")

My students took the problem home for a weekend. All had done well in high school geometry, but they had never dealt with anything akin to hollow tubes and winding string. None returned with an answer at the subsequent class.

On arriving they found, at the center of our table, supplies they hadn't seen since first-grade projects: scissors, tape, string, rulers, plus ten cardboard tubes, which I had been saving from paper towel rolls. Here's what followed:

- The students, working in pairs, were told they should use a ruler to divide the tube lengthwise into four equal sections and designate them by inscribing three circles around the tube.
- One student in each pair was then given an indeterminate length of string, and told to wind it symmetrically around the tube, while the partner would snugly affix it with pieces of tape. Now it looked like the tube depicted in the problem.

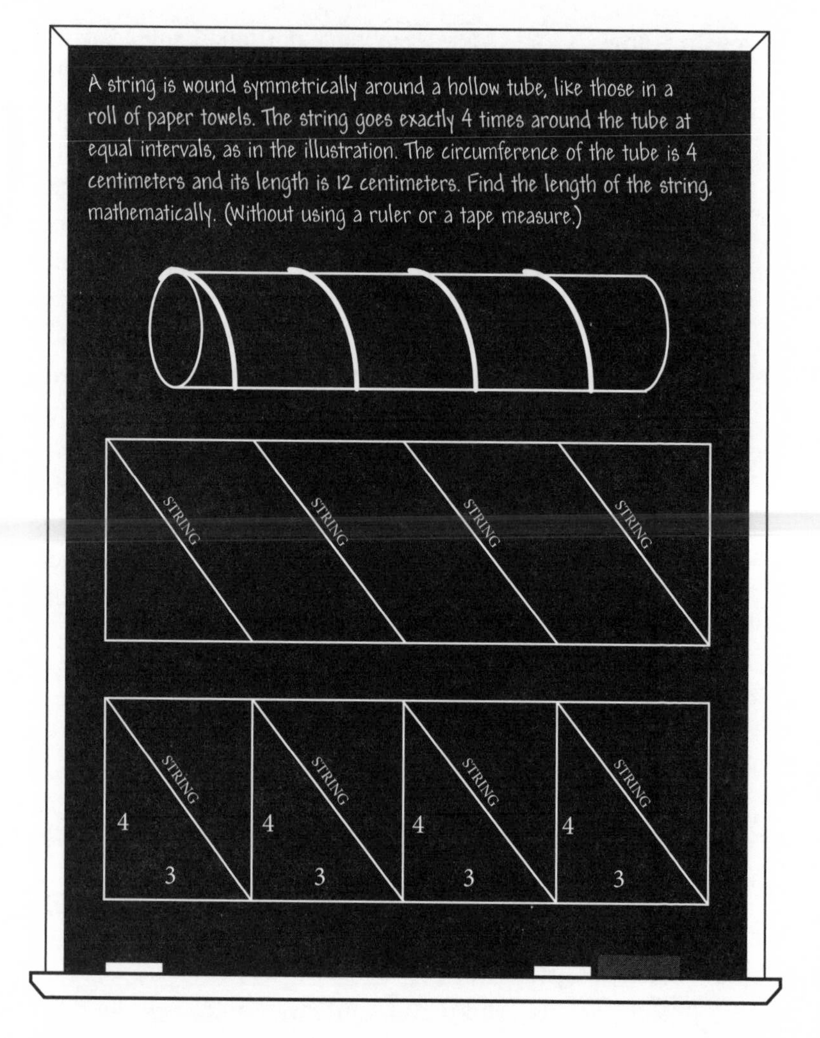

- Next, they were told to use the scissors to cut the tube open lengthwise, and flatten it out with the sections of taped string facing up. As can be seen, the flattened tube shows four diagonal pieces of string. Regard them as long sides of four right-angle triangles.
- The lengths of the other two sides of the triangles are easily obtained. One comes from what we were told is

> the circumference of the tube. The other is one-fourth of the tube's length. So we need only seventh-grade Pythagoras to find the lengths of the diagonals, which added together supply the answer.

Cutting the tube was a necessary step toward a solution. Granted, the instructions didn't specify this. But they *did not* say that that *could not* be done. Rather than delving deeper into geometry, the answer needed scissors, string, and tape. So what's wanted are tests that call for reflection, imagination, and a willingness to think outside the box—or, in this case, the tube. Or tests that don't expect a single predetermined answer, but acknowledge that there can be several suitable solutions to a problem—which is generally the case in the real world. So what's being asked of mathematics is that it show some flexibility. But then again, that's not a trait usually associated with this discipline.

VOICES

As a university-level biologist, I have never had occasion to prove that two triangles are congruent. We were told that if we could think logically about triangles, we could think logically about all sorts of things. What nonsense!

Unlike literature, history, politics, and music, mathematics has little relevance to everyday life. And I say this as a professor of mathematics.

From ten years of teaching high school math, I can only ask if it is necessary that every student know how to simplify a rational expression or to rearrange a quadratic equation?

Since math departments have failed on a colossal level to get their point across to lots of students, we need an alternative way to teach these dropouts.

VOICES

As a mathematician, I believe that the burden is on the advocates of requiring algebra to show that it gives a greater benefit than studies it competes with. It's a claim that needs to be backed up by a strong argument—or, better yet, empirical data.

In my forty years as a math educator, I have seen too many capable people crippled by this algebra curse. We are losing too many students to the gatekeepers who push algebra as some kind of miracle drug for success in life.

As a teacher of math courses ranging from arithmetic to calculus, I suggest that programming, statistics, and finance are better uses of most students' time. Algebra is not the only way to teach disciplined thinking.

As a middle-school teacher, I will have close to two hundred students who now believe they are failures because they did not meet excessive math standards.

11

How Not to Treat Statistics

Most of us are intrigued by statistics, at least when they're presented in a friendly form. That is why *USA Today*'s front page regales its readers with vignettes like these:

> Where American pet dogs sleep: 42% on owners' beds; 33% on bedroom floors; 22% in another room; 3% outside.
>
> Principal causes of debt: house plus car, 25%; credit cards, 21%; college loans, 12%; medical costs, 12%.
>
> Favorite condiments: ketchup, 47%; mustard, 22%; mayonnaise, 20%; salsa and others, 11%.
>
> Average life spans of paper money: $1 bills, 18 months; $5 bills, two years; $10 bills, three years.

Less frivolously, these statistics appeared in an issue of the *New York Times*'s Sunday Review:

> 17.2% (health care share of 2012 gross domestic product)
>
> Over 44,000 (articles published on obesity in 2013)

$9.39 and $7.25 (real value of minimum wage in 1968 and 2013)

Up to 27,000 (ski jobs lost to lower snowfall, 1999–2010)

For his wise and witty bestseller, John Allen Paulos used the terms *numeracy* and *innumeracy.* He chose these plays on *literacy* to describe a comparable set of skills he felt too many people lack. The Mathematical Association of America prefers *quantitative literacy,* which has a more academic ring but comes to much the same thing. Harvard's faculty, never wishing to be outdone, says it wants all its graduates to be adept at *quantitative reasoning,* hinting at a higher mental plane. I'll be examining this aim and claim later on.

In fact, people can be quite nimble with numbers, at least when they involve subjects close to home or that otherwise attract their interest. Sports fans analyze players' "stats," parse the odds, take part in betting pools, and draw up budgets for fantasy teams. Others fill in Sudoku grids or factor coupons for the family budget. In my numeracy classes, I've seen students with varied aptitudes become agile with statistics. Nor should this be surprising. Numbers are a language, and all of us learn its rudiments early on. Still, if figures are used for counting, we need to be clear about what's being tallied. You can't analyze economic inequality meaningfully unless you know how wealth and income are measured and which sources are reliable. I'll be showing how this can be done.

HOW INNUMERATE ARE AMERICANS?

A 2013 *New York Times* editorial sought to alarm its readers by stating that "only 18 percent of American adults can calculate how much a carpet will cost if they know the size of the room and the per square yard price of the carpet." If this is so, and assuming we're dealing only with straight dimensions, it's truly disgraceful that 82 percent couldn't compute the cost correctly. In our era, most adults have completed high school, with at least some geometry and algebra. Yet we're being told they can't even apply the arithmetic they learned in elementary grades.

But before joining the lament, let's look at the question that spurred the accusation. It's from the most recent U.S. Department of Education survey of adult literacy, which also covers quantitative skills. The centerpiece of the problem is an advertisement, reproduced here along with the question as it was worded.

It soon emerges that what's being tested is as much literacy as numeracy, as real-life situations often are. In this case, it starts

with a verbal feint. You must stop to wonder whether you should reduce the announced price of $9.49 by 41 percent, which would bring it down to $5.60 per square yard.

It's best that you don't. The 41 percent "taken off" has already been factored in, since its earlier price was $15.99. Evidently, at least some people made this misstep, contributing to the 82 percent error rate. The next ploy is wilier. Note that the dimensions for the floor area you want covered are given in linear *feet*. Yet the carpet's price is given per square *yards*. So it's only a first step to say the area of the carpet measuring nine by twelve feet works out to 108 square feet. You then must have the insight to transform those 108 square feet into 12 square yards, which also requires recalling how many square feet there are in a square yard. (The correct answer is 12 yards × $9.49/yard = $113.88.)

The high error rate on the carpet question should lead us to ask *why* many adults got the answer wrong. In fact, what it reveals is that any effort to improve quantitative reasoning must stress careful reading as well as instilling computational skills. At the same time, the question we've examined calls for only elementary-grade arithmetic, not any application of higher mathematics.

In the very same test, 52 percent of the sampled adults were able to compute a discount on a heating oil delivery, and 69 percent correctly interpreted a graph showing income disparities. But these were straightforward questions, with no gimmicks, tricks, or unit conversions. Even so, success rates of 52 percent and 69 percent warrant two cheers at best and certainly give rise to concerns.

My contention is that time and effort that might have been devoted to nurturing numerical agility has been given over to asymptotes, rational exponents, and other esoteric topics that only those who choose to major in mathematics in college will ever encounter again. The carpet problem is evidence that the logic ostensibly learned during years of geometry and algebra didn't help these adults to distinguish square feet from square yards. This failure calls to mind a remark by Deborah Hughes-Hallett, a professor of mathematics at the University of Arizona: "Advanced training in mathematics," she reminds us, "does not necessarily ensure high levels of quantitative literacy."

EQUATIONS UNDER THE HOOD

For this chapter, I will be focusing on two sorts of statistics. I'll call one kind *public statistics* and the other *academic statistics.* The first refers to numbers that impinge on our lives, either personally or in our social experience, like those at the opening of this chapter, which told us student loans account for 21 percent of personal debt, or that healthcare was more than 17 percent of GDP in 2012. To understand and use public statistics requires only sixth-grade arithmetic and a skeptical eye for suspicious sources and ideological bias.

I'll be using the term *academic statistics* to describe a scholastic discipline increasingly taught in high schools and colleges, and continuing in postgraduate seminars and academic research. Academic statistics is a domain of teachers, professors, and scholars, who are devoted to their methodologies and have made their pursuit a respected career. Most had their initial training in mathematics, which they regard as integral, indeed indispensable, to what they study and teach. Whether for reasons of scholarship or status, the overwhelming majority of those pursuing academic statistics have no sympathy for modes of analysis that simply need arithmetic.

The following—which I've slightly abridged—was written by Edward Frenkel, a mathematics professor at the University of California's Berkeley campus, shortly after an article of mine appeared. (For the record, I supported retaining algebra, but proposed that alternatives also be offered.)

> In a recent op-ed, Andrew Hacker suggested eliminating algebra from the school curriculum and instead teach how the Consumer Price Index is computed.

> What seems to be completely lost on Hacker is that the calculation of the CPI is in fact a difficult mathematical problem which requires deep knowledge of all major branches of mathematics including advanced algebra.

So indulge me for a moment while I show how I used the Consumer Price Index with my own students. I took my cue from Richard Scheaffer of the University of Florida, a former president of the American Statistical Association. "The key to statistical thinking," he has said, "is in the context of a real problem and how data might be collected and analyzed to help solve that problem."

Comparing CPI findings for 1994 and 2012, my students and I focused on how and where households in those years were spending their money. We looked at forty-five categories, ranging from fresh fruits and vegetables (down 36 percent) to "personal care" (up 118 percent). The goal was to deploy numbers to analyze personal and social changes. In our class discussions, a ground rule was that you had to cite figures to support your interpretations. Thus, on one side they found less spending on alcoholic beverages (down 40 percent). On another front, there was a rise in mostly sugary soft drinks (up 18 percent). Now that we knew the figures, how to explain these shifts?

As we saw, Frenkel sees the CPI in an entirely different light. What interests him are the complex equations used to create the data in the first place and to keep it up to date. Thus in his article, he printed out:

$$\mathbf{PC_{annual} = [(IX_{t+m} / IX_t)^{12/m} - 1] * 100}$$

This equation, he explained, is used to factor in inflation rates for items ranging from frozen foods to home mortgages. In his view, one can't begin to use the Consumer Price Index without first absorbing enough algebra to parse equations like this one.

Here's the issue. Frenkel is talking about the mathematics used to set up the CPI and keep it updated. While I acknowledge and appreciate those efforts, my aim is to use CPI's figures to sharpen our understanding of ourselves and our society. I cannot see what's gained in making undergraduates master advanced equations before they can start discussing why Pepsi-Cola is replacing Budweiser. No one would say we can't use our laptops without first studying the chips and circuits behind the screen.

IS YOUR PHONE BILL TOO HIGH?

I attended the 2013 meeting of the National Council of Teachers of Mathematics, held that year in Philadelphia. It was a large and buoyant gathering, where faculty members from high schools and colleges exchanged ideas and promoted innovations. One spirited session was on Quantitative Financial Literacy, featuring a recently created course at a suburban New York high school. It was billed as "the perfect third or fourth year course for everybody!" While pressure is on to have all students take a fourth year of mathematics, it's become evident that at least some can't cope with calculus. Hence this alternative, announced as having lots of "real world applications," such as savings account interest, withholding tax rates, and depreciation on a car. This sounded great. But of course the devil is in the details. So let's look at one of its "real world" lessons (see the illustration on the following page).

I didn't know whether to laugh or cry. Or both, if that is possible. Can educators believe that telephone owners, youthful or otherwise, will construct equations like the ones on the next page to check their charges? At the session's end, I asked the presenters why this and all their other "financial literacy" lessons called for at least intermediate algebra. "We had to," one of them told me. "Only then would the course be approved as a mathematics offering." What we've seen here is how academic statistics not only impresses its stamp, but thwarts using arithmetic-based methods, which would be just as effective—and a lot more applicable—for public numeracy.

Cell Phone Expenses

A phone calling plan has a basic charge per month, which includes a certain amount of free minutes. There is a charge for each additional minute.

The split function below gives the price $f(x)$ of an x-minute phone call.

Fractions of a minute are charged as if they were a full minute.

$$F(x) = \begin{cases} 40 \text{ if } x \leq 750 \\ 40 + 0.35(x - 750) \text{ if } x > 750 \text{ and } x \text{ is an integer} \\ 40 + 0.35([x - 750] + 1) \text{ if } x > 750 \text{ and } x \text{ is not an integer} \end{cases}$$

Describe the cost of the plan by interpreting the split function.

ADVANCING PLACEMENT

We turn to schools with the hope that they will transform young people into knowledgeable and thoughtful adults. Along with safe driving and safe sex, statistics is being added to curriculums. In 2013, fully 169,508 pupils took advanced placement courses in the subject, nearly triple the 58,230 a decade earlier in 2003. (Statistics about statistics!) As this growth continues, it might be argued we are on course for creating a statistically sophisticated citizenry.

On first hearing, this could seem consonant with what I've been proposing. So I decided to look further and arranged to attend several AP statistics classes at a well-regarded high school near my college. All AP courses use a common curriculum, designed by the College Board, which retains panels of professors to specify the syllabus.

The reason for the uniformity is that a uniform examination is given in each AP subject at the end of the semester. The presumed purpose of these tests is to tell colleges which entering students are qualified to go straight on to advanced courses. Another is to allow applicants to impress admissions officers with an ambitious transcript. Here is a typical AP examination question on statistics:

In a test of Ho: μ = H versus Ha: $\mu \neq 8$, a sample size of 220 leads to a p-value of 0.034.

Possible Answers: Which of the Following Must Be True?

(a) A 95% confidence interval for μ calculated from these data will not include $\mu = 8$.

(b) At the 5% level if *H* is rejected the probability of a Type II error is 0.034.

(c) The 95% confidence interval for μ calculated from these data will be centered at $\mu = 8$.

(d) The null hypothesis should not be rejected at the 5% level.

(e) The sample size is insufficient to draw a conclusion with 95% confidence.

Correct Answer: (a)

Given this way of approaching statistics, perhaps it's not surprising that, of the 169,508 students who took the 2013 test, only 58 percent obtained passing grades. And we should recall that this is an optional class, intended for students aspiring to competitive colleges. So it's in order to ask what is ensured by installing a national program with so high a failure rate. A common answer is that setting a high bar betokens a commitment to stringent standards. Of course, there are plenty of contests that end with very few winners, something young people already know. But here we're talking about schools, which we ask to educate their pupils.

If nothing else, those who see nothing wrong with a 42 percent failure rate should let the rest of us know their reasoning. This issue will reappear when decisions are made about the mathematics scoring system in the Common Core.

A FOUNDATION FALTERS

The Carnegie Foundation for the Advancement of Teaching was alarmed. Its officials had found that virtually all of the 1,191 community colleges in the United States have a singular requirement. Incoming students must obtain specified scores on standardized mathematics tests before they can start taking courses for credit. Carnegie's researchers discovered that fully 60 percent of would-be community college entrants didn't meet the bar and so were assigned to remedial sections. Even worse, 80 percent either failed these noncredit classes or subsequently failed when they took a regular mathematics course. Either way, not passing algebra meant they couldn't start college.

So Carnegie selected nineteen community colleges, in states ranging from Washington to Florida, and offered them a deal. The colleges were asked to overlook the mathematics scores of some of their entrants and exempt them from remedial sections. Instead, they would be allowed to take a class Carnegie had created, called Statways, which it described as an introductory statistics course. Funding for full-time instructors would be provided, with the classes no larger than twenty students. During two academic years, 2011 through 2013, a total of 1,817 students were enrolled in this project.

When Statways was announced in 2010, I was heartened and impressed. In a *Chronicle of Higher Education* article, its creators affirmed their desire to "help solve the remedial-math problem" that takes so heavy a toll. They said they had designed "a statistics pathway that will provide a challenging alternative," showing students how "statistical reasoning" could be "an essential aspect of their everyday lives." These words were what I had long hoped to hear: statistics for citizens. So I looked forward to Statways' results.

Four years later, Carnegie released a report on Statways. Perhaps the most telling statistic of its own was that of the 1,817 students, only 920—almost exactly half—had passed with a grade of C or higher. The others got Ds or Fs, or dropped during the semester. True, this is somewhat better than the overall failure rates in remedial mathematics. But the Carnegie experiment was closely monitored, with small classes and professional support. What happened?

The answer wasn't hard to find. For reasons not given, Carnegie chose to align its syllabus with a document called *Guidelines for Assessment and Instruction in Statistics Education.* These protocols, which fill two volumes, are promulgated by the American Statistical Association, whose leading members are university-level professors. Some of its tenets seem sensible, like "using real data of interest to students is a good way to engage them in thinking about relevant statistical concepts." After all, "real data" is what ordinarily distinguishes statistics from formal mathematics. But it turned out that these remarks are only the tip of the *Guidelines* iceberg. Its main concern is to strengthen the domain of academic statistics. Here is a sampling of the subjects that Carnegie's community college freshmen were expected to master:

Chi-Square Test for Homogeneity in Two-Way Tables
Special Properties of the Least-Squares Regression
Line Multiple Representations of Exponential Models

In a word, Carnegie's alternative for struggling students ended up as a standard academics statistics course, as promulgated by a panel of research professors: the mandarins of statistics. It remains to wonder why the foundation settled for so orthodox a path. Given its shielded status outside the academy, it's dismaying that Carnegie embraced the rubrics of a self-referential discipline, rather than devising templates of its own for teaching statistics to potential dropouts. Equally disappointing was its unwillingness to urge community colleges to abandon their fatal mathematics hurdles—*that* would have been an interesting experiment. After all, Carnegie itself had figures showing their ruinous consequences. In the end, the foundation settled for launching what it billed as a lifeboat, although only half of its occupants survived.

HARVARD: TALK OR WALK?

Several years ago, Harvard University decided—at least on paper—that it wanted all of its students to graduate with a grounding in *quantitative reasoning*. I like this phrasing. It seems akin to quantitative literacy, but with a cerebral edge. So I took a look at how Harvard's faculty implemented a promising idea.

Harvard was not the first top-tier institution to install a cross-disciplinary "big picture" field of study. Columbia University's undergraduate college has long had a course, which all its students must take, called Contemporary Civilization. Since its syllabus cuts across subject lines, faculties from across the campus participate. On a given morning, one might encounter a professor from economics leading a discussion of *Macbeth*. Across the hall, a psychologist would be parsing the Punic Wars. One of the premises of the Columbia course was that its subject—civilization—could be taught by anyone accredited in the arts and sciences. In a similar vein, Amherst College had its Evolution of the Earth and Man, which was required of all sophomores. It was taught by faculty from astronomy, geology, and biology, who rotated lectures and led discussion sections. Their collaboration exposed students to a shared scientific culture.

These might seem some possible formats for a Harvard-wide course on Quantitative Reasoning. If it followed Columbia's and Amherst's paths, it would have a common framework and benefit from insights of various disciplines. Unfortunately, any such proposal would face a constraint. At Harvard, once professors have full status, they cannot be compelled to take on tasks they find uncongenial. This sheltering is often seen as vital to academic freedom. While a majority of the faculty formally voted for the quantitative requirement, it soon emerged that few had any desire to assist in its teaching.

There would be nothing like a single QR 201 that all undergraduates would have to take. Instead, in a recent year I sampled, a total of fifty-three courses were listed as satisfying the QR requirement. By my count, forty-four of these were existing listings, most long offered by departments. Over half were in mathematics, including courses like Honors Abstract Algebra and Multivariate Calculus, despite the fact that these fields have no palpable connection with quantitative literacy or reasoning.

Preexisting classes in computer science, engineering, and physics were also allowed. Economics, government, and sociology added the methodological courses they already had in place for their own majors. But a closer look showed these courses to be highly mathematical and geared to specialized research, far removed from the kinds of numbers most Harvard graduates would encounter in their personal and public lives. Other professors refurbished offerings they were already giving, like deductive logic and computer programming.

By my count, members of the Harvard faculty created only five new courses that seem to capture the spirit of quantitative reasoning. One, by a professor of statistics, was called "Your Chance for Happiness (or Misery)" and explored the numbers behind our becoming rich or poor, lonely or loved, frustrated or satisfied. Another, by a professor of astronomy, admittedly more specialized, dealt with "The Visual Display of Quantitative Information," which showcased the interplay of aesthetics and accuracy. A philosophy professor created "What Are the Odds?" which dealt with issues like risk, statistical inference, and how correlation differs from causation. A team representing the fields of philosophy, linguistics, and computer science collaborated on "Making Sense: Language, Thought, and Logic." The fifth, taught by a computer scientist, was deftly called "Bits: Information as Quantity, Resource, and Property." While none of these five courses claimed to cover the whole quantitative scene, they reflected the personal passions of the professors, which always makes for good teaching.

Still, the fact remains that these were only five out of the fifty-three QR offerings, with the rest being conventional courses, with or without new titles. The upshot at Harvard is that trying to install a truly new vision of quantitative reasoning will be an uphill battle. The chief problem, notably at research universities, is that professors get embedded in their turfs, which is where professional careers are made and reputations are burnished. To create a course that isn't tied to an established discipline calls for a self-confidence not all academics have. Even holders of tenured chairs worry lest they be charged with popularizing or oversimplifying, not to mention being unscholarly. Imagine the chatter at the faculty club on hearing that a colleague was assigning Darryl Huff's cheerful classic *How to Lie with Statistics.*

All told it's not surprising that of Harvard's 1,334 full-time faculty, only thirty-eight decided to try their hand at a quantitative course. Of the ninety-four members of its mathematics department, only two chose to join the QR team. And doubtless concerned about their colleagues' opinions, they made algebra a prerequisite for their offerings. Attitudes like these help to explain why a survey by the *Crimson,* the student newspaper, found the quantitative requirement to be "the Harvard humanities students' biggest nightmare."

THE PRICE OF PRESTIGE

As with mathematics, professors and teachers of statistics see it as their mission to maintain the purity of their discipline. Hence their penchant for keeping their subject exceedingly arcane, on a plane far from the statistics encountered by even the general run of college graduates. In their view, if statistics is to be taught, it must always be at this rarefied level.

For my part, I admire academic statistics and rely on its equations in much of my work. Thus I have drawn on Gini Ratios to compare income distributions (Norway gets .260 versus .410 for the United States). And I've recently used the Pearson Coefficient to compare mathematics scores of countries with their infant survival rates. (It came to -0.093, which means no correlation at all.) But I don't compute these formulas myself. Rather, I type the numbers into a readily available website and watch the results pop out. (Warning: you had better be familiar with the data, to gauge whether what comes out seems reasonable.)

If we want more people to be versatile with what I've called public statistics, we won't attain that goal if we first demand that they become expert with least-square regressions and line multiple representations. Rather, let's examine what happens in real life. Suppose a local newspaper prints a table showing how medical expenditures vary across the states. You may find that these figures convey some interesting information. However, it's unlikely you will try to apply the $\mathbf{H_0: \mu = H}$ ***versus*** $\mathbf{H_a}$ that you had to imbibe in an AP statistics course.

So why do statistics mandarins insist on so academic a syllabus, even for high school students? Here we're back to safeguarding status and preserving purity. Were they to admit that only arithmetic is needed to become statistically adept, it could undercut their claim to scholarly standing. The same holds at high schools, which promote academic statistics for a fourth year of mathematics. At either level, the needs of students are being sacrificed to preserve the prestige of their instructors.

12

Numeracy 101

In the fall of 2013, I visited my college's Department of Mathematics with a proposal. Most of our students are required to take an introductory mathematics course. I offered to teach an experimental section, which would focus on quantitative reasoning. I made it clear that my assignments would rely almost entirely on arithmetic, but at a rigorous level and often in ways not ordinarily employed. Its aim would be to make students agile with numbers, including the use and analysis of statistics. So along with being a professor of political science, I added being a professor of mathematics to my résumé. It's been said that at New York's Bellevue Hospital, the interns practice on their patients. In that spirit, I want to thank my students in Mathematics 110 for the trials and errors they amiably endured. What follows is a sampling of what we covered.

WORDS VERSUS NUMBERS

I started by chalking two words on the board.

In order to think, communicate, and otherwise express ourselves, we find it useful, if not necessary, to employ both words and numbers. So I asked: when do we use which? At first glance, the answer may seem obvious. Or is it? In fact, we seldom stop to ask what can words do that numbers can't, and vice versa.

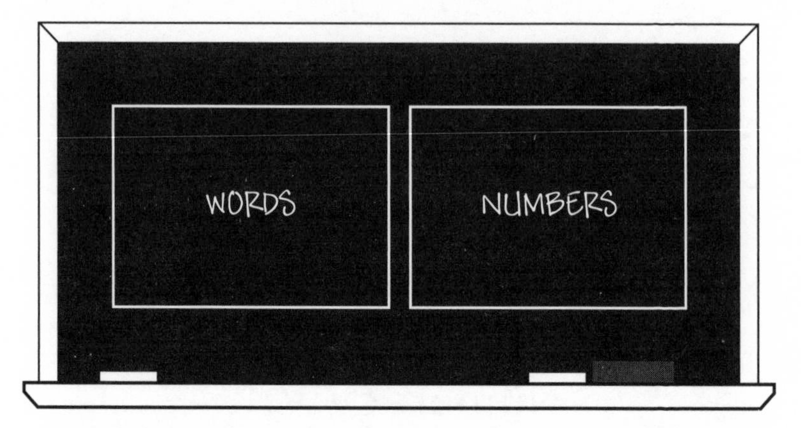

Imagine you hear a friend of yours exclaim, "I have a horrific fever!" Does this tell you more than *103.6°*? (And we haven't even brought in matters of cadence and tone.)

Or here are some words we frequently use: *some*, *few*, *many*, *most*, *several*. Let's look at *most*. Formally, we might agree that *most* ranges from 50.1 percent to 99.9 percent. But toward the lower end, we're more likely to say *about half*, and use *almost all* when we get toward the top. So I asked the members of the class to write down where, on the 50.1 to 99.9 span, they thought *most* might start and where it should stop. Not surprisingly, there was no consensus. Nor were people able to persuade one another, which was also revealing.

We can agree that numbers impart precision—as with 103.6—while words allow for subtlety, intricacy, and nuance. Here are three favorite words of mine: *sedulous*, *meticulous*, *punctilious*. True, they are similar. But all three of them are needed, because they have distinctive shadings, and may be used in different settings. A chef may be called meticulous, while an accountant is punctilious. While a number like, say, 61.8 may seem stark and unadorned, it can bring varied interpretations depending on how it is used. Such as: "only 61.8 percent" or "a whopping 61.8 tons."

Another question: when and how does a number become a

statistic? We all count all the time. While recently waiting on line at my local post office, I counted that there were eight people ahead of me. But I wouldn't call that a statistic, unless it were meant for a study of post office service. Generally, statistics are compiled for some purpose, even when they are spewed out in "raw" form, as is common with pages of undigested data in official publications. But before statistics can be compiled, they have to be created. They aren't just there, waiting to be gathered up, like walnuts under a tree.

I scanned our classroom for a minute, then did a little pecking at my calculator and proceeded to write *31.6 percent* on the board. We now have a statistic, I announced, albeit to puzzled stares. So I told them that this represents the number of people in this room who are visibly wearing glasses. (It was six of nineteen, including myself. Given the small total, I should have rounded it to *32 percent.*) Whether 31.6 or 32, creating figures is an advance over settling for saying that *some* people in the room are wearing glasses. We agreed that precision is preferable to ambiguity. Or at least it is when precision is possible.

Other issues arise with statistics. We should ask where they come from, how they were assembled, why they are being gathered, and how far they can be trusted. In our classroom, a count did the job and could be readily checked. But suppose we wished to know how many Queens College students altogether wear glasses? I would be addressing issues of sampling and probability later on.

DECIMALIZING TIME

Perhaps the chief use of numbers is to *measure.* Nowadays, we quantify just about everything, from heights and weights to traffic accidents and amounts of money. Next, we looked at the efficiency of various methods of measuring. On one side is the metric system, where all calibrations are functions of the number ten. Among its weights are grams, milligrams, centigrams, and kilograms. With metric lengths, there are exactly a thousand meters in a kilometer. Other systems wander all over the place. Inches and feet and yards, for example, have no rational relation to one another. Nor do ounces and pounds, or quarts and gallons. Perhaps the most chaotic are seconds, minutes, hours, and days. To highlight that, I asked my class to tell me, as quickly as they could, what is 27 percent of a day. If needed, they could use pencil, paper, and a calculator. But I wanted the 27 percent stated precisely, in hours, minutes, and seconds. My students have been telling time for most of their lives. Yet several minutes went by before anyone had an answer, and not all were correct.* Finding 27 percent of an entity as familiar as a day turns out to take quite a bit of effort.

As has been noted, there's an alternative, which most of the rest of the world uses: the metric (or decimal) system. But even its adherents have not applied it to time, which means they'd have as much trouble with the 27 percent question as we do. So in the class, I decided to try something that no one has recently proposed: to decimalize time.

Imagine we were asked to prepare a report proposing an entirely new system for our clocks and calendars, all to be based on variants of ten, along with a commentary on how these changes might affect our lives. We might start by recasting minutes so that they have 100 seconds and put 100 minutes in every hour. After that we would replace our current 24 hours with a ten-hour day, and then move to a ten-day week and a ten-month year.

The first question might be: are these shifts physically possible? We know that nature gives us interludes of light and darkness, which we call night and day. Even so, there's nothing in the natural order which necessitates 60-minute hours and months with 28 or 30 or 31 days. The intervals we now use came about by hap-

---

*Well, 27 percent of 24 hours works out to 6.48 hours. What is .48 of an hour, in minutes and seconds? Since an hour has 3,600 seconds, .48 of that is 1,728 seconds, or 28.8 minutes. So the answer ends up as 6 hours, 28 minutes, and 48 seconds.

penstance, in some long-ago eras. A ten-day week isn't any less "natural" than one with seven days. But it's also clear that we'd have to organize our lives differently with a ten-day cycle.

The table on the next page shows a decimal day. Its 100-minute hours would be considerably longer than we use now, having the equivalent of 144 of our current minutes. But need that affect how we conduct our lives?

A ten-day week can also spark discussions. Would we want a three-day weekend, after seven working days? Our current two-day weekend comes to 28.6 percent of the week. Three days off per ten would be 30 percent, which isn't a huge change. As we noted, the natural order doesn't care how many days we put in a week. The class agreed that there shouldn't be eight straight days of work or school. But a lengthy three-day weekend didn't stir much support either. So someone suggested a single day off in the middle of the work-school interval. We took off a little time to conjure names for the three added days. Like "Funday" for the midweek respite.

Is a ten-month year possible? Here too the answer is yes, simply by alternating 36-day and 37-day months, which still yields 365 days. This would seem simpler than our current method of 30 days, 31 days, and 28 or 29 days. Nor does our present system accord with lunar cycles. Full moons don't come on or near the same date each month. It was hard to see what might be lost were months a few days longer. And here's a plus: recall our effort in computing 27 percent of our current week. Since a decimal week has precisely 100 hours, 27 percent would simply be 27 hours. No pencil, paper, or calculator needed.

HOW MANY DAYS IN A YEAR?

Suppose we wanted to decimalize the number of days in a year. Perhaps with 1,000 short days, or 100 long ones, or trying a 500-day year. All that was needed to decimalize seconds through months was some ingenuity, along with a willingness to do things

OUR CURRENT DAY	A DECIMALIZED DAY
24 Hours	10 Hours
1,400 Minutes	1,000 Minutes
86,400 Seconds	100,000 Seconds
Midnight (AM/PM = Meridian)	Midnight (AD = Decimal)
1 AM	
2 AM	
	1 AD (2:24 AM)
3 AM	
4 AM	
5 AM	2 AD (4:48 AM)
6 AM	
7 AM	
	3 AD (7:12 AM)
8 AM	
9 AM	
10 AM	4 AD (9:36 AM)
11 AM	
Noon	5 AD (Noon)
1 PM	
2 PM	
	6 AD (2:24 PM)
3 PM	
4 PM	
5 PM	7 AD (4:48 PM)
6 PM	
7 PM	
	8 AD (7:12 PM)
8 PM	
9 PM	
10 PM	9 AD (9:36 PM)
11 PM	

differently. But when we come to years, ingenuity isn't enough. Our next assignment was to understand how years differ from minutes, days, and weeks.

Galileo famously said, "The Great Book of Nature Is Written in Mathematics." Many of nature's laws have a mathematical basis. Newton, Kepler, and Einstein replicated them in elegant equations. Along with phenomena like earthquakes and cyclones, nature also has some numbers that control or explain how the world works.* One of them is π, whose 3.14159 goes on infinitely, at least so far as we know. Finding it was a great breakthrough, since it equipped humans to measure the area of a circle or the volume of a sphere. And here's another of nature's numbers: 365.2422. It denotes a period of time that governs a basic functioning of the natural world: This is exactly how many days it takes for the Earth to circle the Sun. (For our analysis, four decimal places will suffice.)

This time span is an implacable natural fact, which humans have no choice but to accept. Nor can 365.2422 be readily translated into a decimalized figure like 100.0000. So we humans have had to use and work with nature's 365-day year. The tricky part, of course, is that extra .2422 of a day.

To embark on showing how we have attuned ourselves to nature, I have reproduced a portion of the front page of two issues of a newspaper from over a century ago, as shown on the next page. Examine them carefully, to see if something is a bit odd. The papers are clearly dated, successively, Wednesday, February 28, and Thursday, March 1. And the year is 1900. For a while, my class couldn't see what was being asked of them. So I tried to help, asking them to think about what *isn't* there. Then came an "aha!" moment: there was no February 29 in between. Could it be that 1900 wasn't a leap year? After all, 2000 definitely was.

With the leap years we're accustomed to, inserting February 29 adds a quarter of a day to each year in a four-year sequence. So

*Mathematicians use "natural numbers" for an abstract class, alongside rational numbers, imaginary numbers, even amicable numbers.

ONE CENT. SUNDAY, THREE

The New York Times.

NEW YORK, WEDNESDAY, FEBRUARY 28, 1900.—FOURTEEN PAGES.

ONE CENT. SUNDAY, THREE

The New York Times.

NEW YORK, THURSDAY, MARCH 1, 1900—FOURTEEN PAGES.

taking four years as a group, each year averages having 356.2500 days. But this puts us slightly ahead of nature's 365.2422, to which we must conform. What most people do *not* know is that we deal with this by omitting February 29 three times in every four centuries. We can recall that there was a February 29 in 2000. But there wasn't one in the years 1700, 1800, and 1900, and there won't be one in 2100, 2200, and 2300. So humans have the power to define some elements of time—seconds, minutes, hours along with days, weeks, and months—any way we wish. But the number of days in a year is set by an external universe, whose rules we must identify and obey.

HOW LARGE IS WEST VIRGINIA?

According to the *Statistical Abstract of the United States*, the full area of West Virginia—including lakes and its share of rivers—is 24,230 square miles. But look at the map: most of its boundaries follow meandering paths, usually due to those rivers. So how can so irregular an area be measured so precisely? (Adding up the sizes of the counties won't do it, since many of them are just as erratic.) I should add that 24,230 has been the official figure for at least a century and was ascertained long before satellite photographs and electronic equations.

The maps included here show how it can be done. The trick is to compare the area within the boundary of West Virginia with the full area of the rectangular page on which the map is printed.

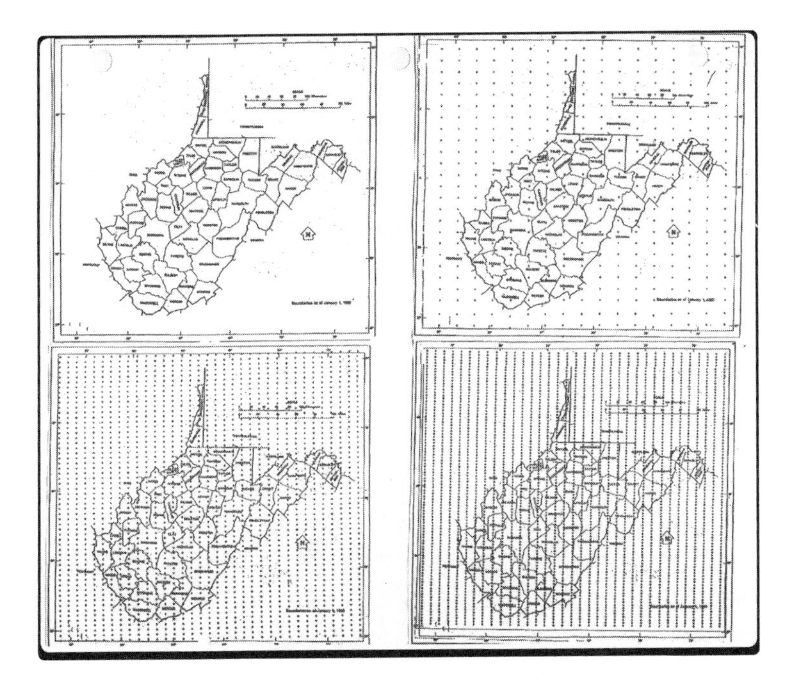

Using the scale on the map, transpose its inches into miles, and from that, compute the number of square miles on the full rectangular page. File that away. The next step will be to find what percentage of the page's square-mile total falls inside the boundaries of West Virginia itself.

Start by calling up the symbols on your word-processing program, and find the bullet dots: • • •. Fill a sheet the size of the rectangular map page with regularly spaced dots and multiply out how many dots you have on the page. After that, superimpose your dot page over the map page, so the entire page looks sprayed with dots. (Doing this will take a bit of trial and error with a photocopier.) The next step will be much more labor intensive.

I asked my students to count up how many dots were within West Virginia itself, including those falling on its borders. The counts they came up with were quite close and we settled for an average. They then calculated the percentage that the within-state dots were of the total of full-sheet dots. Since they knew the number of square miles in the full sheet, the in-state's dot percentage should approximate West Virginia's area in square miles relative to that of the full sheet. With the dots as spaced in the previous paragraph, there were 450 in the full rectangle and 119 within West Virginia, yielding an estimate of 26,444 square miles, or 8.8 percent off the official figure.

We then worked with another map, whose dots were closer together, much like • • • •. This counted out to 1,995 dots on the full page and averaged 528 within West Virginia, which translated to 25,132 square miles, or 3.7 percent off. A third map evoked a lot of groans about the impending count, since there was hardly any space between the dots, which looked much like ••••••. With them, the counts were 3,255 for the page, and averaged 866 within the state, which rendered into 24,018 square miles, or 0.9 percent below the official figure.

The more dots that are squeezed onto the map, the closer they come to covering the entire state and to offering a more accurate estimate of its true area. It's a bit like using calculus to compute the area under a curve. Calculations for square miles only visualize the area of West Virginia as flat, as if it were a totally level plane. In fact, the state is extremely mountainous. So, to keep things interesting I closed the lesson by asking for thoughts on how to compute its three-dimensional area.

VOTES CAST VERSUS SEATS WON

In 2012, a total of 5,556,330 Pennsylvania citizens, spread out among its eighteen congressional districts, cast ballots to elect the state's members in the U.S. House of Representatives. The final tally showed that 50.3 percent of those voters voted for Democratic candidates, while 48.7 percent went to Republicans. (The other 1.6 percent went to other parties.)

Yet when the votes were compared with the actual outcome, it turned out that Republican candidates had won thirteen—or 72 percent—of the eighteen seats. The question for the class: how did the votes that Republicans cast accomplish this, especially since they accounted for less than half of the total balloting?

The sheet on the next page, reproduced from the website of the Clerk of the House of Representatives, was given to the class. What it provides is basically "raw" data, showing the votes recorded in the districts, without any accompanying interpretations or explanations. All that's there are forty numerical totals: the votes for thirty-six major candidates and four minor contenders. So the class had to compute percentage figures for each race. With the numbers and percentages in hand, the assignment was to explain how the Republicans secured so many seats and the Democrats got so few. Students were also given a map of the state showing the district lines and were told to read the article on "gerrymandering" in Wikipedia.

Our class worksheets revealed that the thirteen victorious Republicans averaged 59 percent of the votes cast in their districts. In other words, they won with comfortable, but not overwhelming, margins. However, the five Democratic winners averaged 76 percent, which is far more than they needed to win. So large numbers of Democratic votes were "wasted." That is, they could have helped other of their party's candidates if they had been cast elsewhere in the state.

Two more statistics underscore this imbalance. Statewide, Republicans averaged 181,474 votes in the seats they won, while Democrats averaged 271,970, which was not only a lot higher, but again many more than they needed. Still another count shows that fully 87 percent of Republican voters ended up supporting a winner. Among Democrats, only 49 percent did.

52

PENNSYLVANIA

For Presidential Electors

Republican	2,680,434
Democratic	2,990,274
Libertarian	49,991
Green	21,341

For United States Senator

Tom Smith, Republican	2,509,132
Robert P. Casey, Jr., Democrat	3,021,364
Rayburn Douglas Smith, Libertarian	96,926

For United States Representative

1.	John Featherman, Republican	41,708
	Robert A. Brady, Democrat	235,394
2.	Robert Allen Mansfield, Jr., Republican	33,381
	Chaka Fattah, Democrat	318,176
	James Foster, Independent	4,829
3.	Mike Kelly, Republican	165,826
	Missa Eaton, Democrat	123,933
	Steven Porter, Independent	12,755
4.	Scott Perry, Republican	181,603
	Harry Perkinson, Democrat	104,643
	Wayne W. Wolff, Independent	11,524
	Michael Bryant Koffenberger, Libertarian	6,210
5.	Glenn Thompson, Republican	177,740
	Charles Dumas, Democrat	104,725
6.	Jim Gerlach, Republican	191,725
	Manan M. Trivedi, Democrat	143,803
7.	Patrick Meehan, Republican	209,942
	George Badey, Democrat	143,509
8.	Michael G. Fitzpatrick, Republican	199,379
	Kathy Boockvar, Democrat	152,859
9.	Bill Shuster, Republican	169,177
	Karen Ramsburg, Democrat	105,128
10.	Tom Marino, Republican	179,563
	Philip Scollo, Democrat	94,227
11.	Lou Barletta, Republican	166,967
	Gene Stilp, Democrat	118,231
12.	Keith J. Rothfus, Republican	175,352
	Mark S. Critz, Democrat	163,589
13.	Joseph James Rooney, Republican	93,918
	Allyson Y. Schwartz, Democrat	209,901
14.	Hans Lessmann, Republican	75,702
	Michael F. Doyle, Democrat	251,932
15.	Charles W. Dent, Republican	168,960
	Rick Daugherty, Democrat	128,764
16.	Joseph R. Pitts, Republican	156,192
	Aryanna C. Strader, Democrat	111,185
	John A. Murphy, Independent	12,250
	James F. Bednarski, Bednarski for Congress	5,154
17.	Laureen A. Cummings, Republican	106,208
	Matthew A. Cartwright, Democrat	161,393
18.	Tim Murphy, Republican	216,727
	Larry Maggi, Democrat	122,146

As hardly needs saying, it was a Republican legislature, elected in 2010, that drew the districts that ensured these results. In theory and on paper, every Pennsylvanian is supposed to get an equal vote. But in practice, gerrymandering uses basic arithmetic to make some votes have a greater impact than others.

DISCOVERING PI

Pi is one of the most amazing discoveries in human history.

Note that I said it was *discovered*, not invented or created by some human beings. Pi is another of nature's numbers, which have always existed in their own right. In other words, pi has always been there, much like an earthly element, waiting to be uncovered. In fact, the ancient Babylonians came pretty close, estimating it at 3.125, while early Egyptians got it as far as 3.16049. Knowing pi empowers us to carry out tasks we could not otherwise perform, at least with its singular simplicity. It allows us to measure rounded spaces, like the area within a circle, or the inside volume of an orange, or the outer surface of a conical hat. After all, none of these volumes or areas can be calculated by a tape measure or a ruler.

I asked my students to see whether we could discover pi ourselves. Our method would be an exercise in *reverse engineering*. This is what Lexus does when it goes out and buys a new Mercedes-Benz, and then delicately takes it apart to determine how its competitor's car was put together. Using reverse numeracy, we started with a solution, and then worked back to see how and why it came out as it did.

As we know, pi can tell us the circumference of a circle: $c = \pi d$. Since we were acting as if we did not know the actual digits of pi, we could find them by working with both the circumference and diameter of a circle. Since pi is being cast as unknown, our equation becomes $\pi = c/d$.

The first step is to visit a local bakery and ask for one of those flat cardboard dishes on which layer cakes are placed. It will probably be about ten inches across. Next, cut a piece of string long enough to reach around the rim of the circle. Then carefully run the string around the perimeter, affixing it firmly with tape, every inch or so. After that, cut the string precisely where it completes the circle. Now untape the string and measure its length with a ruler. This is our estimate of the circle's circumference.

After that, measure the cardboard circle's diameter and divide the number you get into the circumference. Does this produce something close to 3.14? In our class, some students were somewhat off, going as high as 3.19. (The trick is to tape the string very firmly and tightly.) Since you started the assignment by assuming that you didn't know the numbers for pi, you can now say that you have discovered them yourself, or at least came quite close.

VERIFYING PI

It's not as easy to deal with pi when we have three-dimensional objects. For example, I have on my desk a soup can, whose label says that its contents weigh 11¼ ounces or 319 grams. We could presumably confirm this by emptying whatever is in the can onto a scale and noting the weight.

But suppose we wanted to know the internal *volume* of the can, expressed in cubic inches or cubic centimeters. So we resort to pi. Our rulers do tell us that our can's circular top has a radius of 2.15 inches and that its height is 3.75 inches. Pi's formula for cylinders is $\pi r^2 h$. So we square the 2.15-inch radius of the can's top, multiply that product by its height of 3.75 inches, and then multiply that result by pi's 3.14, which altogether yields 54.4 cubic inches. It's a precise enough number. But we might ask why we should we trust it, since it depends on an enigmatic pi, which never bothers to explain how it is able to find a cubic volume of a circular container.

So here's what we did. Each student was told to bring an opened and emptied can to class. They were also asked to look around their homes and find a square or rectangular box, which could be made of plastic or wood or metal, but had to be tightly constructed. For my part, I brought several rulers, along with some pitchers, filled with a dark liquid. The pitchers were then poured into all of the student's cans, filling them to their brims. Since we were interested only in the *spatial* volume of the can, the next step was to transfer the liquid in the cans into more readily measured containers.

The students were asked to pour the contents of their cans into the boxes they had brought. They were then told to use a ruler to find the length and width of each box. That done, they inserted the ruler straight down into the liquid to gauge how far it reached up in the box, or its depth. The three dimensions were multiplied to yield a cubic figure, which could be compared with what pi had produced. It won't be spoiling the story to say that the outcomes were very close. Still, it remains to ask whether our can-to-box exercise in fact "proved" the reliability of pi, or if it was just another clue in a mystery we may never solve.

CATCHING TAX CHEATS

Each year, some nine million Americans file a Schedule C to accompany their regular income tax return. They do this because they operate businesses of their own, which the Internal Revenue Service calls "sole proprietorships." On Schedule C, they are instructed to report how much money they received for the goods or services they had sold during the past year. Equally important, they are told to report the expenses they incurred in operating their business, like outlays for office supplies or attorney's fees.

The IRS leaves it to each proprietor to declare how much they paid out for various purposes. It is essentially an honor system. Expenses are regarded as a cost of doing business, so the more you say those were, the less in taxes you will owe. But it is possible that some people will overstate their expenses to keep their tax payments down. That's cheating, and it's against the law.

The IRS has two main ways to catch tax cheats. For the first, it has worked out expected ranges for various kinds of outlays. So if you claimed that you laid out what look like unusual sums for truck repairs or business lunches, those figures would be flagged as off the scale. But most would-be cheats are aware of these algorithms and try to keep overstatements from attracting attention.

So the IRS deploys a second method, one about which most people have never heard. Take a look at these made-up expense lists that might have been filed by two businesses:

SCHEDULE C (Form 1040) — Department of the Treasury, Internal Revenue Service (99)

Profit or Loss From Business (Sole Proprietorship)

▶ Information about Schedule C and its separate instructions is at *www.irs.gov/schedulec*.
▶ Attach to Form 1040, 1040NR, or 1041; partnerships generally must file Form 1065.

OMB No. 1545-0074 — 2014 — Attachment Sequence No. 09

Part II Expenses. Enter expenses for business use of your home **only** on line 30.

No.	Item	Line	Amount	¢	No.	Item	Line	Amount	¢
8	Advertising	8	934	14	18	Office expense (see instructions)	18	8,189	22
9	Car and truck expenses (see instructions)	9	1,999	36	19	Pension and profit-sharing plans	19	992	16
					20	Rent or lease (see instructions):			
10	Commissions and fees	10	8,772	28	a	Vehicles, machinery, and equipment	20a	755	80
11	Contract labor (see instructions)	11	7,991	00	b	Other business property	20b	744	03
12	Depletion	12	8,426	98	21	Repairs and maintenance	21	809	77
13	Depreciation and section 179 expense deduction (not included in Part III) (see instructions)	13	5,827	25	22	Supplies (not included in Part III)	22	2,023	87
					23	Taxes and licenses	23	677	35
					24	Travel, meals, and entertainment:			
14	Employee benefit programs (other than on line 19)	14	3,008	44	a	Travel	24a	2,099	66
15	Insurance (other than health)	15	888	98	b	Deductible meals and entertainment (see instructions)	24b	990	20
16	Interest:				25	Utilities	25	703	55
a	Mortgage (paid to banks, etc.)	16a	729	88	26	Wages (less employment credits)	26	56,343	15
b	Other	16b	163	74	27a	Other expenses (from line 48)	27a	4,355	88
17	Legal and professional services	17	6,449	45	b	Reserved for future use	27b	6,117	52

SCHEDULE C (Form 1040) — Department of the Treasury, Internal Revenue Service (99)

Profit or Loss From Business (Sole Proprietorship)

▶ Information about Schedule C and its separate instructions is at *www.irs.gov/schedulec*.
▶ Attach to Form 1040, 1040NR, or 1041; partnerships generally must file Form 1065.

OMB No. 1545-0074 — 2014 — Attachment Sequence No. 09

Part II Expenses. Enter expenses for business use of your home **only** on line 30.

No.	Item	Line	Amount	¢	No.	Item	Line	Amount	¢
8	Advertising	8	2344	19	18	Office expense (see instructions)	18	4,130	84
9	Car and truck expenses (see instructions)	9	3,999	16	19	Pension and profit-sharing plans	19	3,998	03
					20	Rent or lease (see instructions):			
10	Commissions and fees	10	10,772	34	a	Vehicles, machinery, and equipment	20a	228	90
11	Contract labor (see instructions)	11	5331	00	b	Other business property	20b	116	43
12	Depletion	12	2,776	92	21	Repairs and maintenance	21	1,009	98
13	Depreciation and section 179 expense deduction (not included in Part III) (see instructions)	13	4,887	10	22	Supplies (not included in Part III)	22	12,027	88
					23	Taxes and licenses	23	667	91
					24	Travel, meals, and entertainment:			
14	Employee benefit programs (other than on line 19)	14	32,009	77	a	Travel	24a	2,083	33
15	Insurance (other than health)	15	884	16	b	Deductible meals and entertainment (see instructions)	24b	2,990	13
16	Interest:				25	Utilities	25	7,003	92
a	Mortgage (paid to banks, etc.)	16a	927	22	26	Wages (less employment credits)	26	66,023	15
b	Other	16b	153	12	27a	Other expenses (from line 48)	27a	5,334	26
17	Legal and professional services	17	17,444	35	b	Reserved for future use	27b	19,884	22

Now I'll tell you that one set is an honest report, for which the business has valid receipts. In the other, the numbers are ones that the cheating proprietor made up, trying to make them seem reasonable and realistic. Can you detect which is which?

The IRS thinks it can make a good try. It starts by presuming that cheaters haven't heard of Benford's Law. However, the IRS has and has factored it into the software that reviews tax returns. The law is named after Frank Benford (1883–1948), a General Electric physicist who discovered it almost seventy years ago. He found that in examining any set of real-world numbers, the frequencies of their initial digits follow predictable patterns.

As a result, we can estimate how many initial ones or fours or sevens to expect in a list that gives the lengths of the world's longest rivers, or the populations of large cities, even stock prices at the close of a day. The box below shows the distribution of initial digits in a typical list with 1,000 entries (zeros aren't counted). As can be seen, the odds of a seven or an eight or a nine appearing are considerably less than a one or a two or a three. To be sure, real-world lists won't have precisely these distributions; that's asking too much. Still, it's a fair certainty that ones, twos, and threes taken together will appear much more often than sevens, eights, and nines.

Benford's Numbers

Ones	301	Fours	97	Sevens	58
Twos	176	Fives	79	Eights	51
Threes	125	Sixes	67	Nines	46

Benford's Law has also been found to apply to the way digits appear on honest tax forms, where taxpayers have entered actual numbers, as reflected in receipts they have on hand. But cheaters tend to pull their numbers out of the air, even if they keep them within reasonable ranges. And they are as apt to begin their figures with a seven as they are with a two. That's a revealing mistake: sevens are considerably less common in the actual world. By this time, you can see which is the dishonest Schedule C. The IRS calls in such filers and asks them to document those alleged business dinners and attorney's fees.

WHAT WE BUY AND WHAT IT TELLS US

Throughout the year, the Bureau of Labor Statistics surveys a substantial sample of Americans to find out what they are buying and how much they paid for each item. This *consumer expenditures survey*—as it is called—has several purposes. One is to trace changes in prices, which are compressed into the *consumer price index*, our official gauge of inflation. A second is to keep tabs on how people spend their money. For instance, between 1973 and 2013, household expenditures for food were found to have dropped by 33.2 percent, while outlays for education rose by 69.2 percent. Statistics like these encapsulate important information. While everyone is vaguely aware that tuitions have been rising, the CES tells us exactly how much we are paying out.

Like all public agencies, the Bureau of Labor Statistics churns out volumes of raw data, which it makes available for analysis. (That is how the percentages in the preceding paragraph were obtained.) But it also selects figures that it thinks should be getting attention. Hence, in 2013, it sorted purchasing patterns by income levels. In this case, it divided the nation's households into five equally sized classes. So if we liked, we could compare the spending patterns of the poorest fifth of households with those of the best-off 20 percent.

The table on the next page was our basic document. As can be seen, the poorest fifth devotes 56.3 percent of its income to food and housing, whereas the top fifth needs only 42.4 percent for those basics. As a result, the top quintile is able to spend a larger

Purchasing Patterns: High and Low

Households	Poorest 20%	Richest 20%
Average spending	$22,393	$99,257
Persons per household	1.7	3.2
Earners	0.5	2.1
Vehicles	0.9	2.8
How They Spend What They Have		
Food	16.3%	11.3%
Housing	40.0%	31.1%
Apparel	3.2%	3.1%
Health care	8.0%	5.8%
Personal care	1.2%	1.2%
Entertainment	4.5%	5.4%
Education	3.7%	3.0%
Vehicle purchases	3.8%	6.9%
Alcoholic beverages	0.8%	0.9%
Tobacco products	1.3%	0.3%
Donations	2.6%	4.2%
Retirement	1.6%	14.8%
All other items	14.0%	12.0%

percentage of its available money on cars and other vehicles. If both tiers spend about the same percentages on apparel and alcohol, the poorest commit more to health care and tobacco. (Might those last two be related?)

Of course, people's tastes and traits can't be expressed solely in numbers. So the next step is to ask what's behind the statistics, like pondering the consequences of having to live on $22,393 versus $99,257. But at that point, one student noted that top-quintile households had almost twice as many members as did the bottom fifth. So we took the statistics home, where we converted percentages into actual dollar amounts, and then divided those by 1.7 or 3.2, to obtain per-person outlays. With that done, the $3,650 versus $11,216 per household for food shifted to a less emphatic $2,150 and $3,495 per individual. For those concerned about a country's growing income gap, it helps to know how lacking money affects actual lives.

Other of our exercises, using numbers from public and private agencies, cast further light on economic and social disparities. They also illuminated how far race and gender and national origin correlate with poverty and wealth. Figures from opinion polls help to explain why some conditions persist and also how social attitudes can change. Not everything can be measured by statistics, but they can help to get discussions started.

COMPARING COUNTRIES

Numbers have the advantage of being precise. As we've observed, *72.4 percent* is preferable to *most*, at least if you feel the statistic is reliable. If artfully presented, figures can also tell a story, as we saw with purchasing patterns. They may even prompt moral judgments. Here's an exercise that had such overtones.

Surveys show that most Americans—even if fewer than in the past—view their country as being the best in the world. But can this sentiment be backed by facts? For this exercise, two nations were chosen: Norway and the United States. (Norway is smaller and in some ways more homogeneous, but whether that affects quality of life is open for discussion.) Students were told to use the Internet to dig into statistical compilations from the UN, the

CIA, and the OECD. The table below shows some of the indicators they found. One avenue of discussion was how far factors might connect to one another. For example, could the extent of child poverty predict how many people end up in prison? Or how might obesity rates relate to spending on health care? Since both nations are democracies, where public preferences count, why do the two countries have such different tax rates?

Two Countries

United States		Norway
$29,100	Average household income	$32,000
4.1%	Military spending % of GDP	1.9%
26%	Taxes % of GDP	43%
1,824	Hours worked per year	1,363
33.8%	Adults obese	10.0%
17%	Children living in poverty	3%
765	Vehicles per 1,000 people	494
16%	Health costs % of GDP	9%
12%	Foreign-born population	8%
715	Prison inmates per 100,000	64
106	Vehicle deaths per 1 million	43
.410	Index of income inequality*	.260

*The higher the index, the more inequality
(.000 would mean everyone had the same income)

CHOOSING BETWEEN CHARTS

You've been asked to prepare a report on how many households in the United States have telephones. (Both land and cell count.)

After studying census data on ownership in each state, you have decided to focus on just two: Connecticut and Arkansas. To illustrate your analysis, you want to include a chart. Now you have narrowed your choice to two depictions, choice A and choice B, which are shown below. Which do you think is the best choice?

- Explain why you selected one over the other.
- In your view, is one depiction more accurate or objective than the other? If there is a "bias," how might you describe it?
- Is there a way you want your readers to react?

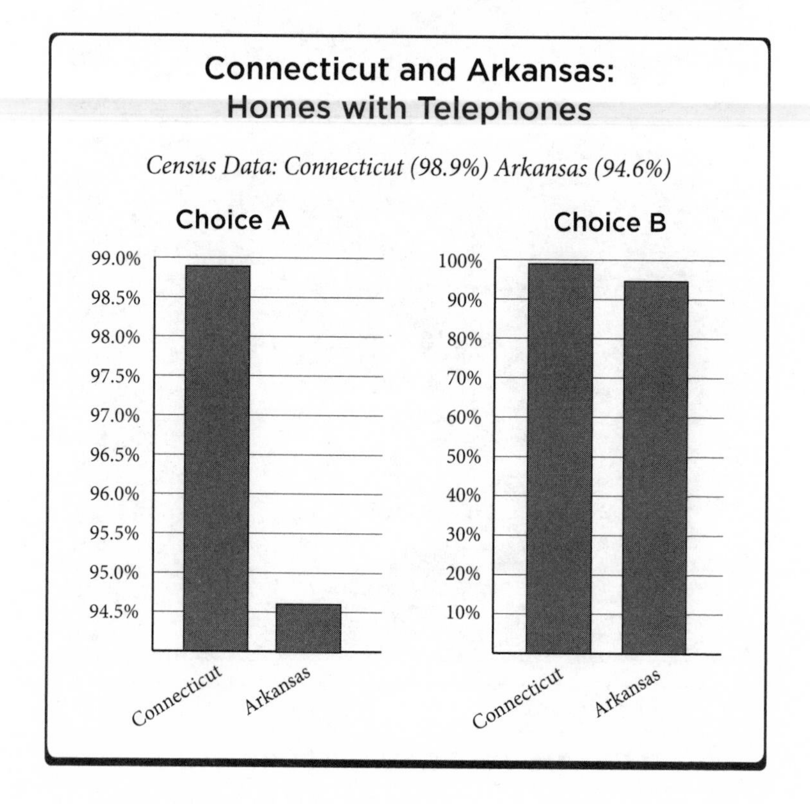

Do you want the difference between 94.6 and 98.9 to come across as large or small? As you see it, is Arkansas's 94.6 fairly close to Connecticut's 98.9, or do you see a significant gulf? Might there be ideological overtones guiding the choices? Numeracy and quantitative reasoning call for certain skills, which this chapter's lessons have addressed. But proficiency with numbers is only part of the story. There's also what you want to do with them and how you want your audiences to react. Here the issues are moral and intellectual, and these too belong in the syllabus.

CODA

My class was offered within a mathematics department, but it could as easily be billed as a language course. By this, I mean that the students were being asked to read, speak, and think *numerically*, as fully as if they were mastering Arabic or Mandarin. In this designation, becoming bilingual leads to interweaving ideas and integers to enlarge our comprehension of ourselves and our world.

Some might argue that young people already have those skills. After all, they learned arithmetic at single-digit ages, followed by algebra and trigonometry in high school. And here we meet one of mathematics' most pervasive myths: that its azimuths and asymptotes are needed to understand actual people and places like Norway and Pennsylvania and West Virginia. (In fact, hardly any mathematics programs allow time for such analysis.) I hope I've shown that exacting arithmetic will suffice, just as it does for most STEM occupations.

This nation has no shortage of problems. But demanding more mathematics of everyone is not a solution. That said, sharpening our numerical skills could help. Like computing the concealed costs of student loans before signing the bottom line. Or calculating how much immigrants contribute to the economy; whether taxes act as a disincentive, or the relative roles of exercise and diet in losing weight. Surrounding us is William James's "blooming, buzzing confusion." We can make better sense of it if we underpin our facts with reliable figures. This also calls for knowing where the numbers come from, how they are derived, and if they have undisclosed agendas. Another troubling truth is that America's public is neither as informed nor as incisive as citizens in many comparable countries. More attention to quantitative acumen would be a fruitful first step.

ACKNOWLEDGMENTS

No matter how original or idiosyncratic, every book has its predecessors and builds on their work. While I take full responsibility for every comma in *The Math Myth*, I want to cite several authors who were pioneers in realms I subsequently chose to explore. Most especially: Anthony Carnevale, Donna Desrochers, Underwood Dudley, David Edwards, Norman Matloff, John Allen Poulos, and Lynn Arthur Steen.

Some ideas in this book appeared earlier in the *New York Times* and the *New York Review of Books*, where I benefited from the expert oversight of Susan Lehman and Robert B. Silvers, who prove how vital editors are to the literary craft. The same tribute holds for my editors at The New Press: Diane Wachtell, Tara Grove, and Maury Botton. They took my rather off-beat proposal seriously, reined me in when I bit off too much, and piloted nebulous drafts into an actual book. A special thanks to Bookbright Media for their help with the charts and book design. Hazel Lichterman and Mary Cox helped with arranging interviews in the book's early days, while Bill Goldstein had a crucial role in giving it structure and shape. Robin Straus—this is my sixth book with her—is far more than an agent. A cherished friend, she kept cheering me on through *The Math Myth's* longer-than-usual life.

Numerous people assisted me with this book, some without realizing that they did. All the time, I was filing away their comments, recollections, and conversations; their impress is on every page. So immense gratitude to: Eileen Artemakis, Chris Artis, James Bach, Nicholson Baker, Shirley Bagwell, Sydney Beveridge, Jonathan Blattmachr, Stephen Blyth, Jo Boaler, Sy Bon, John Bowman, Martin Braun, Peter Braunfield, Doug Brown, Jerry Brown, Shareen Brysac, Arthur Caplan, Sewall Chan, Robin Chotzinoff, Kenneth Chu, Sandi Clarkson, David Coleman, Michael Crow, Ann Davison, Cornelia Dean, Betsy Dexheimer, Eric Dexheimer, Mark Dziatczak, Kathy Eden, Edward Jay Epstein, Rol Fessenden, Whitney Fielding, Leah Finnigan, Gabe Frankl-Kahn, Ellen Frell, Edward Frenkel, Bill Friedland, Joan Friedland, Ester Fuchs, Howard Gardner, Rene Gernard, Wallace Goldberg, Nan Graham, Vartan Gregorian, Trish Hall, Aisha Hassan, David Helfand, Walter Helly, Oren Jarinkes, Harvey Jay, and R. Steven Justice.

Also: Joseph Kanon, Mark Kantrowitz, Jane Karr, Richard Kayne, Ethel Klein, Chuck Kleinhans, Jonathan Kozol, Michael Krasner, Edward Krugman, Sunil Kumar, Julia Lesage, Gloria Levitas, Mitchell Levitas, Richard Levy, Paul Lockhart, Avi Loeb, Alison Lurie, Ian Lustbader, Peter March, John Matzui, Stephen Mazza, Karl Meyer, Paul Miroulig, Mitzi Montoya, Derek Moore, Robert Moses, Michael Mulgrew, James Muyskens, Nel Noddings, Carolyn Toll Oppenheim, Colleen Oppenzato, Dennis Overbye, Evan Picoult, Pradeep Raj, Patricia Rachel, Diane Ravitch, Mark Reimann, Virgil Renzulli, Karen Ristau, Sam Roberts, Betty Rollins, Daniel Rose, Joanna Rose, Caleb Rossiter, Alex Ryba, William Sayle, Robert Schaeffer, Anya Schiffrin, Gavin Schmidt, Robert Schwartz, Joan Silver, Ray Silver, John P. Smith III, Debora Spar, Joseph Stiglitz, Steven Strogatz, Judith Summerfield, Yan Sun, Vijay Sundaram, Michael Teitelbaum, Mark Tucker, John Uglum, Peter Van Olinda, Jacob Vigdor, Tony Wagner, James Weatherall, Carl Wieman, Gene Wilhoit, E.O. Wilson, and Sia Wong.

And most of all, thanks to my family. Claudia Dreifus, wife, companion, lover, kindred writer and teacher, and sympathetic sounding board, her impress is on every line. Also to Ann Gower, my daughter, and her husband, Tim Gower, whose assurance and good humor buoyed me while writing this book, which after all is intended for their generation.

NOTES

1: The "M" in STEM

1 *We are warned*: Michael S. Teitelbaum, *Falling Behind?* (Princeton University Press, 2014), pp. 173–74

2 *The Gathering Storm*: National Academy of Sciences (2006)

2 *Before It's Too Late*: National Commission on Mathematics and Science Teaching for the 21st Century (2000)

2 *Tough Choices or Tough Times*: National Center on Education and the Economy (2007)

2 "*double the number*": Business Roundtable, *Tapping America's Potential* (July 2005)

2 "*one million additional*": President's Council of Advisors on Science and Technology, *Engage to Excel* (February 2012)

3 *shortfall persists*: Eric Hanushek, et al, "Education and Economic Growth," *Education Next*, Spring 2008

3 *62 percent of new jobs*: American Diploma Project, Achieve, Inc. (November 2009)

3 Charles Mackay's book is still in print (Dover, 2003)

4 *an opinion article*: "Is Algebra Necessary?" *New York Times*, July 29, 2012

5 *one in five of our young people*: *Public School Graduates and Dropouts*, National Center for Education Statistics (2013)

6 Peter Braunfeld, interview

8 *third from the bottom:* Numeracy Proficiency Among Adults, *OECD Skills Outlook* (2013)

8 John Allen Paulos, *Innumercy* (Hill and Wang, 1988)

9 "*too much mathematics*": Anthony Carnevale and Donna Desrochers, "The Democratization of Mathematics," in Bernard Madison and Lynn Arthur Steen, *Quantitative Literacy* (National Council on Education and the Disciplines, 2003)

9 "*There is no more*": Paul Lockhart, *A Mathematician's Lament* (Bellevue Literary Press, 2009)

10 Robert Moses and Charles Cobb, *Radical Equations: Civil Rights from Mississippi to the Algebra Project* (Beacon Press, 2001)

10 Eric Cooper, "The Next Revolution in Black Achievement," *Huffington Post* (March 27, 2015)

11 "*math skills are more*": Greg Duncan, *EdSource*, edsource.org/2013

2: A Harsh and Senseless Hurdle

13 *a veterinary technician*: John Merrow, "A Harsh Reality," *New York Times*, April 22, 2007

13 *major in studio art*: Ginia Bellafante, "Community College Students Face a Very Long Road," *New York Times*, October 3, 2014

14 *ranks twenty-second out of thirty*: *Education at a Glance*, OECD Indicators 2013

14 *twelfth amid thirty-two nations*: *Education at a Glance*, OECD Indicators 2013

15 "*students most often fail*": Lynn Arthur Steen, "How Mathematics Counts," *Educational Leadership* (November 2007)

15 "*more than half*": Jo Boaler, *What's Math Got to Do With It?* (Penguin, 2015), p. xxii

15 *As recently as 1982*: *Digest of Education Statistics*, National Center for Education Statistics (2013)

15 "*more students to drop out*": Shirley Bagwell, "Mathematics Education Dialogues," National Council of Teachers of Mathematics, April 2002

15 "*then you've lost them*": Teresa George, "Requiring Algebra II in High School Gains Momentum," *Washington Post*, January 4, 2011

15 "*turn kids off mathematics*": Gerald Bracey, "The Malevolent Tyranny of Algebra," *Education Week*, October 25, 2000

16 *failing ninth-grade algebra*: Robert Balfanz and Nettie Legters, *Locating the Dropout Crisis*, Johns Hopkins Center for Research on the Education of Students Placed at Risk (2004)

16 *installed "exit" exams*: *End-of-Course Exams*, Education Commission of the States (March 2012)

16 "*in the district will fail*": David Silver, Marisa Sanders, and Estola Zarate, *What Factors Predict High School Graduation in the Los Angeles Unified School District*, California Drop-Out Research Project (2008)

17 "*needed to factor trinomials*": Joseph Rosenstein, "Algebra II + All High Schoolers = Overkill," *Newark Star-Ledger*, April 29, 2008

17 *$7 billion for classes*: David Kirp, "Closing the Math Gap for Boys," *New York Times*, February 1, 2015

17 "*With shrewd tutoring*": Colleen Oppenzato, interview

18 *paying for such aid*: *Evaluation of the K–8 Mathematics Program, Pelham Public School District*, Rutgers University Graduate School of Education (June 2012)

19 *the consequences are dismaying*: *Graduates with UC/USC Required Courses*, Statewide Graduation Numbers

19 "*creating artificial barriers*": Anthony Carnevale and Donna Desrochers, "The Democratization of Mathematics," in Bernard Madison and Lynn Arthur Steen, *Quantitative Literacy* (National Council on Education and the Disciplines, 2003)

19 wanted *to proceed*: *Bridging the Higher Education Divide*, Century Foundation (2013)

19 *consigned to remedial*: Paul Tough, "Who Gets to Graduate?" *New York Times Magazine*, May 18, 2014

20 "*three, four, five times*": Debra Blum, "Getting Students Through Remedial Math," *Chronicle of Higher Education*, October 26, 2007

20 *only five percent of them*: "Tennessee Colleges: 70% 'Need' Math Remedial," *Education Week*, February 19, 2014

20 "*placed on algebra pathways*": *Remediation: Higher Education's Bridge to Nowhere*, Complete College America (2012)

20 "*denied a certificate*": *What Does It Really Mean to Be College and Work Ready?* National Center on Education and the Economy (2013)

20 *"as Latin was used"*: Marc Tucker, email

21 *"could be taught in high school"*: Lynn Arthur Steen, in Bernard L. Madison and Lynn Arthur Steen, *Quantitative Literacy* (National Council of Education and the Disciplines, 2003).

21 *college a try*: Bureau of the Census, *Educational Attainment* (2014)

21 *72 percent didn't pass*: *Creating the Conditions for CUNY Students to Succeed*, City University of New York Retention Task Force (2006)

22 *over 265,000 grades*: *High School and Beyond Transcript Study: 1981–1993*, Institute on Postsecondary Education (1999)

22 *when a "student fails a course"*: Suzanne Wilson, *California Dreaming* (Yale University Press, 2003).

22 *"heard of anthropology"*: Kevin Birth, interview

23 *29,610 applications*: *The Best 379 Colleges* (Princeton Review, 2014)

23 *at least 700*: *Barron's Profiles of American Colleges* (2013)

3: Will Plumbers Need Polynomials?

27 *outpacing overall job growth*: *Math Works*, American Diploma Project, Achieve, Inc. (2009)

27 *proficient in algebra*: Thomas Friedman, *The World Is Flat* (Farrar, Straus & Giroux, 2005)

27 *straight into the workforce*: Cathy Seeley, *News Bulletin*, National Council of Teachers of Mathematics (May–June 2005)

27 *Algebra is essential*: "Q&A with Tom Luce," National Math and Science Initiative, *Dallas News*, August 18, 2012

27 *applicants with adequate skills*: Rex Tillerson, "How to Stop the Drop in American Education," *Wall Street Journal*, September 6, 2013

28 *the most reliable source*: Bureau of Labor Statistics, *Occupational Outlook Handbook* (2014)

29 *mathematics-based skills*: Anthony Carnevale, Nichole Smith, and Jeff Strohl, *Help Wanted: Projections of Jobs and Education Requirements Through 2018* (Georgetown University Center on Education and the Workforce, 2010)

30 *"the number of engineers"*: Joseph Stiglitz, interview

31 *program in laser photonics*: "Community Colleges Respond to Demand for STEM Graduates," *Chronicle of Higher Education*, February 15, 2013

31 *laid off 18,000*: "Microsoft to Lay Off Thousands," *New York Times*, July 17, 2014

31 "*deskilling*": Paul Beaudry et al., *The Great Reversal in the Demand for Skilled and Cognitive Tasks,* National Bureau of Economic Research (2013)

32 "*half of all STEM jobs*": Jonathan Rothwell, *The Hidden STEM Economy,* Brookings Institution (June 2013)

34 *count released in*: National Science Board, *Science and Engineering Indicators* (2012)

34 *recent engineering graduates*: "A College Degree Is No Guarantee," Center for Economic Policy and Research (May 2014)

34 "*failing to produce*": *Math Works,* American Diploma Project, Achieve, Inc. (2009)

34 "*scarcity of qualified people*": John Cornyn, quoted in "Texas on the Potomac," *Houston Chronicle*, July 27, 2011

34 "*going to get worse*": Brad Smith, quoted in informationweek.com, September 28, 2012

34 "*its historical preeminence*": *Engage to Excel*, Report to the President (February 2012)

34 *1,051 recent applicants*: "Skills Don't Pay the Bills, *New York Times*, November 20, 2012

35 "*demand for workers*": www.bls.gov/ooh/about/glossary.htm

35 "*a skill shortage*": "Skills Don't Pay the Bills," *New York Times*, November 20, 2012

35 "*at rock-bottom rates*": Ibid.

35 "*this foundational knowledge*": *A National Talent Strategy: Ideas for Securing U.S. Competitiveness and Economic Growth* (Microsoft, 2012)

35 *awarded in computer science*: U.S. Department of Education, *Digest of Education*

36 *making under $41,000*: American Association of Professional Coders, *2011 Salary Survey,* news.aapc.com

36 *creates and installs armor*: "If You've Got the Skills, She's Got the Job," *New York Times*, November 19, 2012

36 "*not a workplace context*": "What It Takes to Make New College Graduates Employable," *New York Times*, June 28, 2013

36 *Engineer III*: "Help Wanted," *New York Times*, October 26, 2014

38 *fully 262,569 holders*: *Characteristics of H.1B Specialty Occupation Workers*, U.S. Citizenship and Immigration Services (September 2012)

38 *middle or late twenties*: Norman Matloff, "The Adverse Impact of Work Visa Programs on Older U.S. Engineers and Programmers," *California Labor and Employment Law Review* (August 2006)

39 *averaged only 57 percent*: Norman Matloff's H-1B Web page at heather.cs.ucdavis.edu

39 "*are more or less indentured*": Ross Eisenbrey, "America's Genius Glut," *New York Times*, February 8, 2013

39 "*ordinary people*": Norman Matloff's H-1B Web page at heather.cs.ucdavis.edu

39 "*fully competent*": *H-1B Visa Program Reforms Are Needed to Minimize the Risks and Costs of Current Program*, General Accounting Office (January 2011)

40 "*wanting to lower wages*": Norman Matloff's H-1B Web page at heather.cs.ucdavis.edu

40 "*a modern automobile plant*": *Foundations for Success*, The National Mathematics Advisory Panel, U.S. Department of Education (2008)

40 "*people other than mathematicians*": Lynn Arthur Steen, "Teaching Mathematics for Tomorrow's World," *Educational Leadership* (September 1989)

42 *applications of basic arithmetic*: Linda Rosen et al., "Quantitative Literacy in the Workplace," in Bernard L. Madison and Lynn Arthur Steen, *Quantitative Literacy* (National Council of Education and the Disciplines, 2003)

42 "*essential for machining*": Mike Snowden, telephone interview

42 "*from school mathematics*": John P. Smith, telephone interview

42 "*prospective employers lack*": Lynn Arthur Steen, "Data, Shapes, Symbols," in Bernard L. Madison and Lynn Arthur Steen, *Quantitative Literacy* (National Council of Education and the Disciplines, 2003)

42 *keep our aircraft safe*: *Math Works*, American Diploma Project, Achieve, Inc. (2009)

4: Does Your Dermatologist Use Calculus

47 *biology was important*: "Premedical Course Requirements," *Contemporary Issues in Medical Education* (September 1998)

47 "*never needed it*": Penny Noyce, "A Concerned Citizen's Perspective," *Mathematics Education Dialogues* (March 1998)

48 "*only arithmetic is needed*": Richard Kayne, interview

48 *showed him a question*: *MCAT Physics and Math Review* (Penguin Random House, 2014)

49 "*levels of abstract mathematics*": Anthony Carnevale and Donna Desrochers, "The Democratization of Mathematics," in Bernard L. Madison and Lynn Arthur Steen, *Quantitative Literacy* (National Council of Education and the Disciplines, 2003)

49 *Chapman-Kolmogorov equations*: Thomas McGannon, *Study Guide and Solutions Manual for Exam P of the Society of Actuaries* (Stipes Publishing, 2007)

49 "*need in their jobs*": Gabe Frankl-Kahn, interview

50 "*do you actually use?*": David Edwards, email

51 *reproduced a question from*: College Board website, sample questions from the Graduate Record Examination

52 "*calculating the tip*": "How Much Math Do MIT Graduates Use?" *MIT Technology Review* (Summer 2003)

52 "*thousands of engineers*": quoted in Padraig McLoughlen, "Is Mathematics Indispensable?" Mathematical Association of America (January 2010)

52 *in his two fields*: Sunil Kumar, interview

52 *no more than eighty*: Consolidated Edison, interviews

53 "*was always simple*": Julie Gainsburg, "The Mathematical Disposition of Structural Engineers," *Journal for Research in Mathematics Education* (November 2007)

53 "*only eighth grade*": David Edwards, interview

53 "*sophisticated rockets*": Mitzi Montoya, interview

54 *most successful scientists*: Edward O. Wilson, "Great Scientists Don't Need Math," *Wall Street Journal*, April 5, 2013

54 "*the frontiers of science*": Tony Chan, "A Time for Change?" in Chris Golde and George Walker, *Envisioning the Future of Doctoral-Education* (Josey-Bass, 2006)

55 "*no connections with reality*": Avi Loed, interview

55 "*general calculus course*": John Matsui, interview

55 "*mathematics less and less*": Carl Wieman, interview

56 *finished only high school*: Joanna Masingila, "Mathematics Practice in Carpet Laying," *Anthropology and Education Quarterly* (March 1994)

57 *observed fourteen handicappers*: Stephen Ceci and Jeffrey Liker, "A Day at the Races," *Journal of Experimental Psychology* (June 1986)

58 "*the top-ten list*": Tony Wagner, *The Global Achievement Gap* (Basic Books, 2010)

58 "*earnings of students*": *Math Works*, American Diploma Project, Achieve, Inc. (2009)

5: Gender Gaps

63 "*there is no difference*": Christiane Nuesslein-Volhard, interview

64 *scanned the transcripts*: Abigail Norfleet James (ed.), *Teaching the Female Brain* (Corwin Press, 2009)

64 *college mathematics courses*: Jesse Rothstein, "College Performance Predictions and the SAT," *Journal of Econometrics* (July–August 2004)

64 "*boys are more active*": Meridith Kimball, "A New Perspective on Women's Math Achievement," *Psychological Bulletin* (March 1989)

64 *doing in their classes*: Stephen Ceci and Wendy Williams, *The Mathematics of Sex: How Biology and Society Conspire to Limit Talented Women and Girls* (Oxford University Press, 2010)

65 "*low expectations*": John Hennessey, Susan Hochfield, and Shirley Tilghman, *Boston Globe*, February 12, 2005

65 "*if the environment is right*": Leonard Sax, "In England, Girls Are Closing Gap," *Wall Street Journal*, March 30, 2005

66 *found girls averaging*: U.S. Department of Education, *Digest of Education Statistics* (2013)

66 *scored girls at 50.2*: U.S. Department of Education, *High School Longitudinal Study* (2012)

66 *scores in its files*: *ACT Profile Report*, Graduating Class 2013.

66 *girls averaged 499*: *College-Bound Seniors*, College Board (2013)

66 *"built for speed"*: Leonard Sax, "In England, Girls Are Closing Gap," *Wall Street Journal*, March 30, 2005

67 *by mental strategies*: Howard Wainer and Linda Steinberg, "Sex Differences in Performance on the Mathematics Section of the Scholastic Aptitude Test," *Harvard Educational Review* (Fall 1992)

67 *even more revealing*: Ann Gallagher, *Sex Differences in Problem-Solving Strategies*, College Board Report 92-2 (1992)

67 *"very, very carefully"*: Sylvia Nasar and David Gruber, "Manifold Destiny," *New Yorker*, August 28, 2006

69 *the citywide pool*: Al Baker, "Girls Excel in the Classroom but Lag in Entry to Eight Elite Schools in the City," *New York Times*, March 22, 2013

70 *Classical Latin*: New York City Department of Education, *Specialized High Schools Student Handbook* (2014–2015)

71 *several hundred donors*: National Merit Scholarship Corporation, *Annual Report, 2012–2013.*

74 *in seventy-nine seconds*: College Board, "What's on the PSAT/NMSQT?" sample questions

75 *even if indirectly*: "About Our Scholars," National Merit Scholarship Corporation, *Annual Report, 2012–2013.*

76 *about two groups*: College Board, *College-Bound Seniors* (2010)

6: Does Mathematics Enhance Our Minds?

81 *"strengthens the mind"*: quoted in David Eugene Smith, *The Teaching of Elementary Mathematics* (London, 1900), p. 171

81 *"live more intelligently"*: *A Sourcebook of Applications of School Mathematics*, National Council of Teachers of Mathematics (1980)

81 *instills procedural fluency*: *Adding It All Up: Helping Children Learn Mathematics*, National Research Council (2001)

82 *"a distinctive kind"*: Alice Crary and W. Stephen Wilson, "The Faulty Logic of the 'Math Wars,'" *New York Times*, June 16, 2013

82 *"training to think"*: Morris Kline, *Why Johnny Can't Add: The Failure of the New Math* (St. Martin's, 1973)

82 *"one mental skill"*: E. D. Hirsch, "Not So Grand a Strategy," *Education Next* (Spring 2003)

83 *"any single mental function"*: E. L. Thorndike, "The Influence of First-Year Latin upon Range of English Vocabulary," *School and Society* (January 20, 1923)

84 *"to be no research"*: Peter Johnson, "Does Algebraic Reasoning Enhance Reasoning in General?" *Notices of the AMS* (October 2012)

84 *"music and art broaden"*: Underwood Dudley, "Is Mathematics Necessary?" National Council of Teachers of Mathematics (March 1998)

87 *where 527 students*: www.imo-official.org/results.aspx

89 *"an impeccable argument"*: Roger Penrose, *The Road to Reality* (Jonathan Cape, 2004)

90 *"blooming, bustling confusion"*: William James, *Principles of Psychology* (Henry Holt, 1890)

91 *"is proved forever"*: Simon Singh, *Fermat's Last Theorem* (Fourth Estate, 1997)

92 *from one of my trials*: Andrew Hacker, "Who Killed Harry Gleason?" *Atlantic Monthly* (December 1974)

94 *"All the mathematics one needs"*: G.V. Ramanathan, "How Much Math Do We Really Need?" *Washington Post*, October 23, 2010

96 *Reasoning mathematically* and *Are mathematicians the best*: Roger C. Schank, "No, Algebra Isn't Necessary," *Washington Post*, October 3, 2012

96 *I know people who can*: Richard Cohen, "Taking on Algebra," *Washington Post*, July 30, 2012

7: The Mandarins

97 *"mathematics power elite"*: Lynn Arthur Steen, *Achieving Quantitative Literacy* (Mathematical Association of America, 2004)

97 *"self-perpetuating priesthood"*: Paul Halmos, "Mathematics as a Creative Art," *Royal Society of Edinburgh Year Book* (1973)

98 *"halfway to first base"*: Betty Rollins, interview

98 *top-down changes*: *Foundations for Success*, U.S. Department of Education (2008)

99 *"by grade six or seven"*: S. Stephen Wilson, "The Common Core Math Standards," *Education Next* (Summer 2012)

100 *"the quality of work"*: *Understanding Student Success*, American Association of Universities and the Pew Charitable Trusts (April 2003)

102 *a Supra-Sputnik*: Morris Kline: *Why Johnny Can't Add: The Failure of the New Math* (St. Martin's, 1973)

103 *leadership role*: University of California, Irvine, department of mathematics, press release, August 4, 2005

104 *the raw numbers*: U.S. Department of Education, *Digest of Education Statistics,* 1971 and 2013

106 Peter March, Peter Braunfeld, Sia Wong, interviews

107 *would be their choice*: College Board, *College-Bound Seniors* (2014)

108 *"bored out of my mind"*: Stephen Montgomery-Smith, quoted in "Colleges More Often Hiring Part-Timers," *St. Louis Post Dispatch,* March 24, 2012

108 *on undergraduate instruction*: Conference Board of the Mathematical Sciences, *AMS Survey of Undergraduate Mathematical Programs* (2012)

109 *less than 1 percent of students*: Clarence Stephens, "A Humanistic Academic Environment for Learning Undergraduate Mathematics," SUNY-Potsdam (undated)

110 *inaccessible to undergraduates*: Clarence Stephens, position paper, department of mathematics, SUNY-Potsdam (undated)

110 *their work to one another*: Keith Devlin and Ian Stewart, quoted in "When Even Mathematicians Don't Understand the Math," *New York Times*, May 25, 2004

112 *only 730 American citizens*: U.S. Department of Education, *Digest of Education Statistics* (2013)

112 *"fewer than 40 percent"*: *Engage to Excel*, Report to the President of the United States (2012)

113 *"consternation erupted"*: "Mathematicians' Central Role in Educating the STEM Workforce," *Notices of the AMS* (October 2012)

114 *"The real reason"*: Justin Baeder, "How Much Math Is Too Much?" *Education Week*, July 31, 2012

115 *"Defenders of the existing"*: Roger C. Schank, "No, Algebra Isn't Necessary," *Washington Post*, October 3, 2012

8: The Common Core: One Size for All

118 *"Need Advanced Math"*: Achieve, Inc., *Math Works*, American Diploma Project (2008)

119 *"the request of governors"*: William Kirwan, Timothy White, and Nancy Zimpher, "Use the Common Core," *Chronicle of Higher Education*, June 20, 2014

119 *85 percent in Mississippi: State High School Exit Exams: 2010–2011 School Year*, Center for Economic Policy (2012)

120 *"common national test"*: Edward Glaeser, "Unfounded Fear of Common Core," *Boston Globe*, June 14, 2013

120 *"norm referenced"*: Jeff Nelhaus of PARCC, interview

121 *thesis on Edmund Spenser*: David Coleman, interview

121 *"didn't write them"*: "Duncan Pushes Back on Attacks on Common Core," U.S. Department of Education press release, June 25, 2013

122 *"a Marine officer"*: Ibid.

124 *"skills are the same"*: Ken Wagner, interview

124 *"have comparable preparation"*: Achieve, Inc., quoted in "Are College and Career Skills Really the Same?" PBS Online, June 14, 2013

124 *Pascal's Triangle*: High School: Number and Quantity Common Core State Standards for Mathematics

124 *unabashed tracking*: see Andrew Hacker and Claudia Dreifus, "Who's Minding the Schools?" *New York Times*, June 9, 2013

125 *"Our system isn't ready"*: Mitchell Chester, quoted in "States Grapple with Setting Common Test-Score Cutoffs," *Education Week*, December 13, 2013

126 *"folks for disaster"*: "What Do Math Educators Think About the Common Core?" *Education Week*, April 24, 2013

126 *"a higher dropout rate"*: "Questions Arise About the Need for Algebra 2 for All," *Education Week*, June 12, 2013

126 *Math for Harvard*: Anthony Carnevale, quoted in Dana Goldstein, "The Schoolmaster," *The Atlantic*, October 2012

126 *"multiple pathways"*: "Common Core in High Schools: New Florida Law Raising Questions," *Education Week*, April 22, 2013

126 *has three diplomas*: "Texas Board Votes to Ax Algebra Graduation Rule," *Education Week*, December 4, 2013

127 *the Hawken School*: "Ohio's Private Schools Fight to Avoid Common Core Exams," *Cleveland Plain Dealer*, May 15, 2014

128 *"crony capitalism"*: "The Common Core Is Crony Capitalism for Computer Companies," *Reason*, July 14, 2014

128 *"rigorous, content-rich"*: Kathleen Porter-Magee and Sol Stern, "The Truth About the Common Core," *National Review Online*, April 3, 2013

128 *the "ugly truth"*: "Jeb Bush Defends Common Core," *Sarasota Herald Tribune*, October 17, 2013

128 *"whatever remains on one hundred eighty?"*: "Gov. Bush Stumped by Math Question," *Washington Post*, July 14, 2004

129 *"the mathematics needed"*: "SAT Makeover Aims to Better Reflect Classroom Learning," *Education Week*, March 7, 2014

9: Discipline Versus Discovery

131 *"becomes fun and games"*: "This Is Math?" *Time*, August 25, 1997

131 *how much money*: "Subtracting the New Math," *Newsweek*, December 15, 1997

133 *"don't have to like it"*: "Repetition and Rap," *New York Times*, June 6, 2001

133 *"Mathematics is hard"*: "The Hardest 'R,'" *National Review*, June 5, 2000

133 *"exemplary"* or *"promising"*: U.S. Department of Education, *Promising Initiatives to Improve Education in Your Community* (February 2000)

134 "*master to automaticity*": "Ten Myths About Mathematics Education," http://www.nychold.com/myths-050504.html

135 "*certain amount of drudgery*": Karsten Stueber, letter to *New York Times*, December 10, 2013

135 "*that is not subjective*": Simon Singh, *Fermat's Enigma* (Walker, 1997)

136 "*Students able to persevere*": Camilla Benbow, "Algebra Is Really Worth the Effort," *Nashville Tennessean*, August 15, 2012

139 "*a community of learners*": *Teaching Handbook for the Interactive Mathematics Program*, www.mathimp.org/resources

140 *in cases of 24*: "A Teacher in the Trenches," *New York Times*, April 12, 2000

142 "*scientists and mathematicians*": Al Cuoco, "Some Worries About Mathematics Education," *The Mathematics Teacher* (March 1995)

143 *the competition is conducted*: International Mathematical Olympiad, www.imo-official.org/results.aspx

145 "*depends on collaboration*": Sylvia Nasar and David Gruber, "Manifold Destiny," *New Yorker*, August 28, 2006

10: Teaching, Tracking, Testing

148 "*a disproportionate share*": *Tough Choices or Tough Times*, National Center on Education and the Economy (2007)

148 *lower than classmates*: *College-Bound Seniors*, "Intended College Major," College Board (2014)

149 "*have a natural talent*": *Mathematics Benchmark Report*, Trends in International Mathematics and Science Study (1999)

150 "*multiple intelligences*": Howard Gardner, *Frames of Mind* (Basic Books, 1983)

151 "*little is known*": U.S Department of Education, *Foundations for Success* (2008)

152 *14,000 high school seniors*: "Dividing Opportunities: Tracking in High School Mathematics," Michigan State University (May 2008)

153 *the full field was 59*: Program for International Student Assessment,Organization for Economic Cooperation and Development (2012)

156 "*there is no tracking*": "Japan: A Study of Sustained Excellence," Organization for Economic Cooperation and Development (2010)

157 "*nighttime sleep deprivation*": "Sleep Patterns and School Performance of Korean Adolescents," *Korean Journal of Pediatrics* (January 2011)

157 *strapping tiny pillows*: Amanda Ripley, *The Smartest Kids in the World* (Simon & Schuster, 2013)

159 *4 percent of Americans*: Edward Rothstein, "It's Not Just the Numbers," *New York Times*, March 9, 1998

162 "*Unlike literature, history*": G.V. Ramanathan, "How Much Math Do We Really Need?" *Washington Post*, October 23, 2010

11: How Not to Treat Statistics

165 *vignettes like these*: *USA Today Snapshots* (Sterling Innovation, 2009)

165 *Less frivolously: New York Times*, February 8, 2014

166 *wise and witty bestseller*: John Allen Paulos, *Innumeracy: Mathematical Illiteracy and Its Consequences* (Hill & Wang, 1988)

166 "*only 18 percent*": "Who Says Math Has to Be Boring?" *New York Times*, December 7, 2013

167 *covers quantitative skills*: *National Assessment of Adult Literacy*, National Center for Education Statistics (2003)

169 "*does not necessarily ensure*": Deborah Hughes-Hallett, "The Role of Mathematics Courses in the Development of Quantitative Literacy," in Bernard Madison and Lynn Arthur Steen, *Quantitative Literacy* (National Council on Education and the Disciplines, 2003)

170 "*including advanced algebra*": Edward Frenkel, "Don't Let Economists and Politicians Hack Your Math," *Slate*, February 8, 2013

170 "*key to statistical*": Richard Scheaffer, "Statistics and Quantitative Literacy," in Bernard Madison and Lynn Arthur Steen, *Quantitative Literacy* (National Council on Education and the Disciplines, 2003)

171 *New York high school*: Rich Saroi and Rob Gerver, *Quantitative Financial Literacy: Advanced Algebra with Financial Applications* (South Western Cengage Learning, undated)

172 *fully 169,508 pupils*: College Board, Advanced Placement (2013)

173 *only 58 percent obtained*: Ibid.

174 *the 1,191 community colleges*: *Community College Pathways: 2012–2013 Descriptive Report*, Carnegie Foundation for the Advancement of Teaching (2013)

174 "*the remedial-math problem*": Anthony Bryk and Uri Treisman, "Make Math a Gateway, Not a Gatekeeper," *Chronicle of Higher Education*, April 23, 2010

175 *align its syllabus*: *Guidelines for Assessment and Instruction in Statistics Education,* American Statistical Association (2005)

178 "*biggest nightmare*": "Pay Lip Service to a Vitally Important Skill," *Harvard Crimson,* September 14, 2006

INDEX

ABOUT THE AUTHOR

Andrew Hacker is the author of ten books, including the *New York Times* bestseller *Two Nations*. He teaches at Queens College and lives in New York City.

PUBLISHING IN THE PUBLIC INTEREST

Thank you for reading this book published by The New Press. The New Press is a nonprofit, public interest publisher. New Press books and authors play a crucial role in sparking conversations about the key political and social issues of our day.

We hope you enjoyed this book and that you will stay in touch with The New Press. Here are a few ways to stay up to date with our books, events, and the issues we cover:

- Sign up at www.thenewpress.com/subscribe to receive updates on New Press authors and issues and to be notified about local events
- Like us on Facebook: www.facebook.com/newpressbooks
- Follow us on Twitter: www.twitter.com/thenewpress

Please consider buying New Press books for yourself; for friends and family; or to donate to schools, libraries, community centers, prison libraries, and other organizations involved with the issues our authors write about.

The New Press is a 501(c)(3) nonprofit organization. You can also support our work with a tax-deductible gift by visiting www.thenewpress.com/donate.